STORMY WEATHER

David Corlett is is an academic and writer. He is the author of *Following Them Home: The Fate of the Returned Asylum Seekers* (Black Inc, 2005) and, with Robert Manne, *Sending Them Home: Refugees and the New Politics of Indifference* (Quarterly Essay 13, Black Inc, 2004). He is an Adjunct Research Fellow at Swinburne University's Institute for Social Research, where the research for this book was carried out, and an Honorary Research Fellow at La Trobe University.

BRIEFINGS

A series of topical books exploring social, political and cultural issues in contemporary Australia

Series editors: Peter Browne and Julian Thomas
Australian Policy Online (www.apo.org.au)
Institute for Social Research, Swinburne University of Technology

Stormy Weather

The Challenge of Climate Change and Displacement

DAVID CORLETT

UNSW PRESS

A UNSW Press book

Published by
University of New South Wales Press Ltd
University of New South Wales
Sydney NSW 2052
AUSTRALIA
www.unswpress.com.au

First published 2008

National Library of Australia
Cataloguing-in-Publication entry

Author: Corlett, David.
Title: Stormy weather: the challenge of climate change and displacement/ David Corlett.
ISBN: 978 1 742230 08 5 (pbk.)
Subjects: Human beings – Effect of climate on.
Climatic changes – Social aspects.
Emigration and immigration.
Dewey Number: 304.25

Cover photograph Full moon over the Funafuti Atoll lagoon.
Peter B. Bennetts / Lonely Planet Images

Printer Ligare

This book is printed on paper using fibre supplied from plantation or sustainably managed forests.

Contents

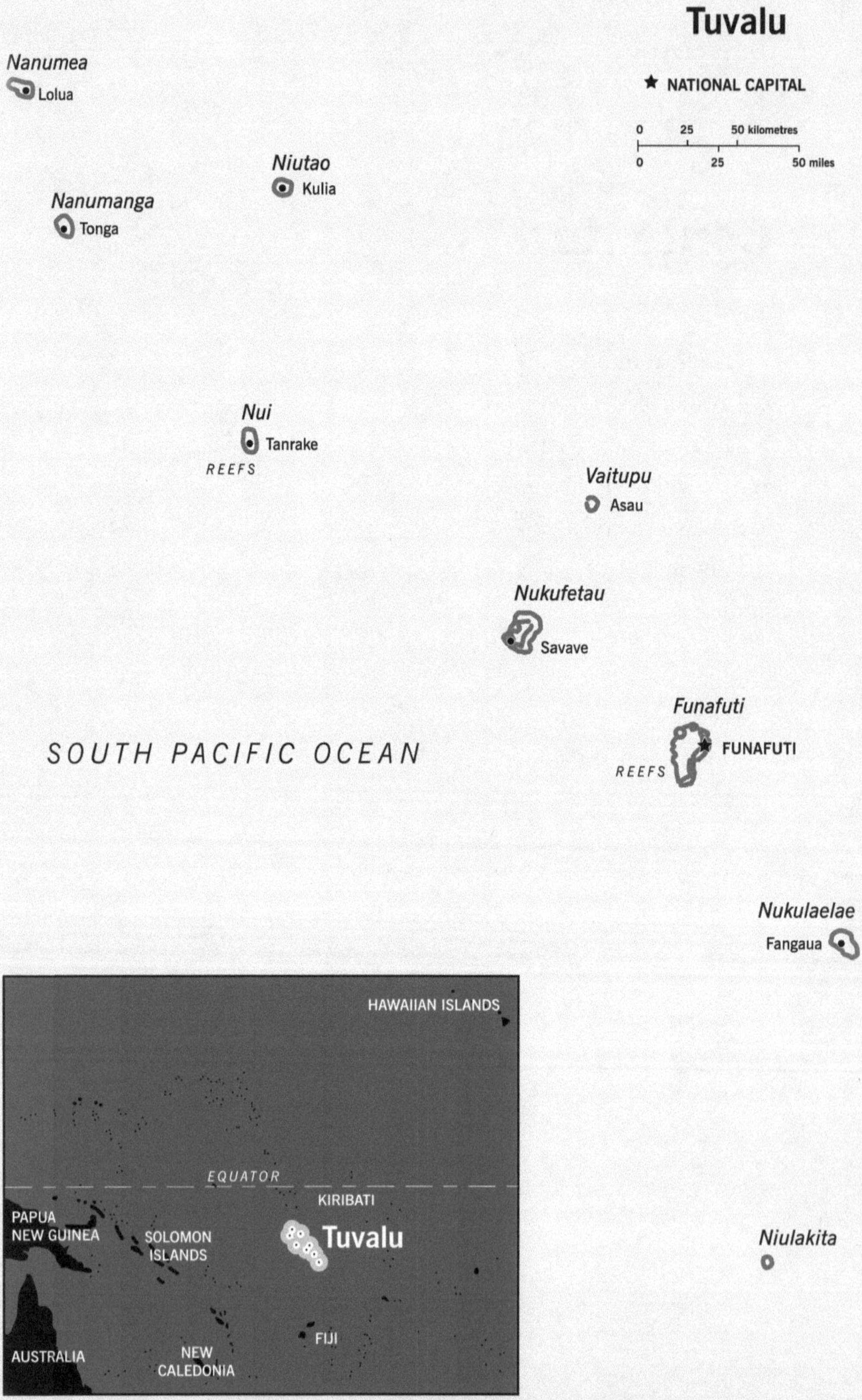
Tuvalu
NATIONAL CAPITAL
0 25 50 kilometres
0 25 50 miles
Nanumea
Lolua
Niutao
Kulia
Nanumanga
Tonga
Nui
Tanrake
REEFS
Vaitupu
Asau
Nukufetau
Savave
Funafuti
FUNAFUTI
REEFS
SOUTH PACIFIC OCEAN
Nukulaelae
Fangaua
Niulakita
HAWAIIAN ISLANDS
EQUATOR
KIRIBATI
PAPUA
NEW GUINEA
SOLOMON
ISLANDS
Tuvalu
FIJI
AUSTRALIA
NEW
CALEDONIA

CHAPTER 1

The great flood

[T]he greatest single impact of climate change could be on human migration – with millions of people displaced by shoreline erosion, coastal flooding, and agricultural disruption.

– Intergovernmental Panel on Climate Change, 1990

In late February 2006, a handful of journalists and a sprinkling of tourists converged on Tuvalu. It was king tide season, but this was a king tide of a different order. Seawater flooded Fongafale, the main islet on Funafuti atoll, the capital of Tuvalu. Water bubbled up, oozing, from underground. Some local residents waded knee-deep in water, others paddled kayaks near their homes. Houses on the lower parts of the islet were inundated. The extent of the flooding was unprecedented.

It was a sensational story. The media saw what it came to find: the new Atlantis. Tuvalu has long been seen as the nation most likely to sink as the sea level rises as a result of climate change. And although Tuvaluans number only eleven thousand, migration from this small Pacific island nation will be part of a much larger movement of people – from Kiribati, Indonesia, China, Bangladesh, and Vietnam, from Southern Africa and the Sahel, and from the countries of South America – caused by climate change–related factors. In the same way that the melting of polar ice has been seen as the "canary in the mine" for climate change, the dislocation of Tuvaluans is viewed as a sign of the broader displacement of potentially hundreds of millions of people as a result of the changes to the Earth's climate.

The especially high king tides of 2006 looked like climate change in action: global warming due to greenhouse gas emissions causing the thermal expansion of the oceans and melting the world's great glaciers. The flooding of Tuvalu, a nation consisting of three reef islands and six coral atolls more than 4000 kms northeast of Sydney, just south of the equator and just west of the international dateline, was considered to be proof that the global environment was in turmoil. Yet it wasn't as straightforward as this.

The media and the other spectators had some time to plan for the king tides of 2006. Two years earlier the South Pacific Sea Level and Climate Monitoring Project had predicted particularly high tides for 2006. The Australian Bureau of Meteorology's National Tide Centre warned that the 2006 king tides would be the highest for the period 1990 to 2016. And rather than sea level rises due to climate change, Fiji's Meteorological Service said that the unusually high tides were due to the coincidence of the closest new moon to Earth for 2006 and high normal tidal factors.[1] That the phenomenon had been foreseen did not, however, make living through the experience any easier. The unprecedented 2006 king tide was alarming. It reinforced Tuvaluans' awareness of their vulnerability as citizens of one of the lowest countries in the world, a country with an average elevation of less than two metres above sea level. It also was consistent with other changes they had been noticing in their natural environment over the past couple of decades.

More than a year later, after the world's media had gone home, Tuvalu's vulnerability was once again apparent. In the hours before dawn on 17 April 2007, Tufitu Lotee's extended family was asleep in their houses in the south of the boomerang-shaped islet of Fongafale. To the east is the vast Pacific Ocean. To the west is the lagoon, a body of water surrounded by the reefs and islets that make up the atoll. Where Tufitu lives, the land between the ocean on one side and the lagoon on the other is no wider than a hundred metres.

For some reason, in the small hours of this particular morning, Tufitu could not sleep. She was reading a book, but she sensed that something was not right. "I heard the sound from the sea," Tufitu told me. "Big waves. So I got up and just went to check up. And then

I saw the waves were just right at the shore, coming up as usual. So I went back. But I couldn't sleep because in my heart I felt that something was going to happen that night. Three times I went to check up. And on the fourth I saw a big wave is coming up. I saw the big waves. I just saw big waves were coming in."

Tufitu lives on the "ocean side" of the islet in a cluster of three houses, one of which is made of traditional pandanus thatch, with her husband, mother, brother, her brother's wife and children, her sister and her cousin. Between the ocean and the homes of Tufitu's family is a bank of old, dead coral, no more than a half a dozen metres wide. There are pigs caged on the bank and a hammock strung between a couple of trees, then the bank falls away and the land flattens out. It is here that the houses begin. When I talked to Tufitu I could see the lagoon across the road between the houses and trees; behind me I could hear the sound of the waves breaking on the reef.

Sitting on an old concrete slab – a reminder of the house that once stood there – Tufitu went on to tell me that when she saw the waves coming early that April morning she went to wake her family. A wave crashed over the bank and the rest of the neighbourhood woke up. One house was hit, according to a report to Tuvalu's Secretary to Government, throwing stones and rubble into the rooms where the family was sleeping.

The sea not only came over the bank; it also seeped out from under the ground. Tufitu has a mark on her house indicating the level to which the water rose. The whole area flooded. According to the government report, written by Sumeo Situ, the disaster coordinator in the Prime Minister's office, the water rose by between 700 and 800 mm. A second set of waves surged over the bank not long after the first.

"The water surrounded us," Tufitu told me. "The water was around us, coming up. And the people, they were helping one another. They were carrying our children over there. And one man came and asked us to take all our children up there, to the high places, to the road, otherwise an electric shock may happen at that time."

A similar wave surge had occurred during daytime in 2002. "But last year," Tufitu said, "it happened in the night. So we were fast asleep and we didn't know what was going to happen. Imagine that: we were all fast asleep and it just happened. And there was no wind when it happened. There was no wind. Only the huge waves that we could hear: boom! boom! boom! I heard it that night. I couldn't sleep that night. What I felt, I really felt frightened. Fear. We were really concerned about the children, and also myself. I just prayed to the Almighty, to help us. Because there is no other place that we can go to."

As daylight broke, the first evacuated family – ten adults and twelve children – was taken to a Red Cross centre in the main village. Another eight families, totalling forty-six adults and sixty-one children, including Tufitu's, soon followed. Some were housed in the Red Cross centre, while others went to stay with other family on the islet or were accommodated in a *falekaupule* or village meeting hall.

The cause or causes of that wave surge are not clear. According to Tauala Katea, scientific officer and acting director of Tuvalu's Meteorological Office, there was no weather system that could have caused the surge. A small earthquake near Tonga may have been the cause. Climate change might also have had an impact. Rising sea levels could have meant that waves that would normally be stopped by the coral bank were able to wash over it. Coastal erosion due to sea level rise might also have been a factor. Tufutu believes it is: "The island is starting to get eaten up. It is getting narrower and narrower and narrower. I believe those people who talk about global warming and weather changing. I believe that."

In response to this emerging threat, locals, with the support of the government and the Red Cross, have started to plant trees to stop the erosion. They are also taking other precautions. "We have started to build up a wall of stones since last year," Tufitu told me. I walked around to the ocean side of the houses, where a small dry wall of coral rocks had been constructed. "We started to build it. But it has started to be eaten up. You can already see some stones have been washed out. It is always changing. The stones and the sand. It is always moving around. I don't know where the sea takes those stones and the sand."

It is difficult to know conclusively whether climate change is one of the factors responsible for the erosion. But if you ask Tuvaluans, many tell you about changes they can see happening in their local environment, changes they say are not normal – changes that are consistent with what we know about the impacts of climate change.

•

The Earth is warming. This is evident in increases in global average temperatures, the melting of ice and snow across the world and the rise of the global average sea level.[2] Eleven of the twelve years from 1995 to 2006 were among the warmest years since instrumental records began in 1850.[3] The world's average surface temperature has increased by 0.6°C over the twentieth century.[4] It is "very likely," according to climate scientists, that the 1990s was the warmest decade and 1998 the warmest year since records started. The increase in temperature in the twentieth century is "likely" to be the highest in the past thousand years.[5]

Global average sea levels rose by an average of 1.8 mm each year between 1961 and 2003 and by an average of 3.3 mm each year between 1993 and 2003.[6] Since 1993, over half of this rise has been caused by the expansion of the ocean due to warming ocean temperatures. Melting glaciers, ice caps and polar ice sheets have also contributed significantly to the rise in sea levels.[7] Further climatic changes consistent with global warming, including changes in precipitation in parts of the world over the twentieth century, have also been evident.[8]

These changes are due in large part to human activity since the Industrial Revolution, most importantly the burning of fossil fuels, which has released greenhouse gases – notably carbon dioxide, or CO_2, but also methane, nitrous oxide and other gases – at historically high levels.[9] These gases remain in the Earth's atmosphere, which, as the term suggests, acts like a greenhouse, trapping heat that would otherwise dissipate, and warming the planet. Last year, the Intergovernmental Panel on Climate Change released its *Fourth Assessment Report*, or *AR4*. Written by 450 lead authors with contributions from a further 800 and peer-reviewed by over 2500 scientists from 132 countries, *AR4* is the most authoritative and conclusive, albeit conservative, assessment of the state of the Earth's climate.[10] In the cautious language of the scientific community, the authors of the section of *AR4* dealing with the physical science basis of climate change concluded that "[m]ost of the observed increase in global average temperatures since the mid-20th century is *very likely* due to the observed increase in anthropogenic greenhouse gas concentrations [emphasis in original]."[11]

As a consequence of the warming climate, natural systems across the globe are changing. Glacial lakes are expanding. Permafrosts are melting, leading to instability and avalanches. The timing of natural processes such as leaf-unfolding, bird migration and egg-laying has changed. Plant and animal species have moved towards the poles and to higher elevations. There have been changes in marine and freshwater systems, affecting fish stocks and the abundance of algae and plankton. There is also evidence that climate change is affecting agricultural production, forestry management and human health. Rises in the sea level, along with other human activities, are contributing to a loss of coastal wetlands and mangroves and increased coastal damage from flooding.[12]

Projections indicate that climate change will have significant future consequences for life on Earth. The Intergovernmental Panel on Climate Change, or IPCC, has developed models that seek to project future emissions and the potential impact on the climate. According to these projections, there is "high agreement" and "much evidence" that global emissions "will continue to grow over the next few decades."[13] If emissions were kept at 2000 levels, temperature increases of 0.2°C per decade for the next two decades and 0.1°C per decade over the rest of the century would be anticipated.[14] Future temperature rises over the same period would depend on the rate of future emissions. Ongoing emissions at present or higher levels will lead to more rapid temperature rises with dramatic consequences.

The ongoing warming of the earth's temperature will lead to rises in the sea level. *AR4* suggests a rise of between 0.18 and 0.59 metres by the end of this century, a decrease on the projected increase in the *Third Assessment Report*. But the sea level figures in *AR4* are extremely rubbery. The authors concede that "[b]ecause understanding of some important effects driving sea level rise is too limited, this report does not assess the likelihood, nor provide a best estimate or an upper bound for sea level rise."[15] With such a caveat, it is difficult not to wonder why the upper limit was assigned in the report at all. NASA's chief climate scientist, James Hansen, has argued that the sea level figures included in *AR4* are too low. Based on the paleoclimatic record – the actual experience of the Earth in the past as told through scientific evidence found in ancient ice, fossils and other sediments – he suggests that if the Earth's temperature rises between two and three degrees Celsius the ice sheets in both Greenland and west Antarctica are likely to collapse, causing sea levels to rise by a metre every twenty years.[16] The temperature range that Hansen uses is within the IPCC's predicted temperature rise of between 2°C and 6°C if greenhouse gases are not reduced. Similarly, the Garnaut Review noted that "some of the recent scientific work suggests that future sea-level rise could be much worse than the sea-level rise outcomes projected in the IPCC's Fourth Assessment Report."[17]

The speculation regarding the sea level is part of a broader uncertainty about what are referred to as non-linear "tipping points." These are the thresholds beyond which changes caused by global warming become, if not irreversible, then extremely difficult to turn round, and have potentially far-reaching "feedback" effects on other parts of the climate system. One example of a tipping point would be the irreversible melting of the Greenland icesheet which, on its own, would cause an increase in sea level of around seven metres,[18] and which could in turn precipitate further atmospheric warming. Already, the Greenland ice sheet is melting faster than previously predicted.[19] It has been suggested that should current rates of greenhouse emissions continue, the melting of the Greenland ice sheet may reach its tipping point during this century.[20]

Even without these non-linear effects, it is predicted that future climate change will have important implications. Ecosystems will be unable to cope with the changes, leading to the loss of plant and animal species. In some parts of the world, agricultural production will decrease; if temperatures increase above 3°C, agricultural production across the globe is projected to decline. Water stress will increase. Some regions, including those that are already dry, will experience a fall in water resources, while others will experience increased precipitation and flooding. Industry and human settlements, particularly those located near the coast and river flood plains, will be affected. Human health will suffer with the spread of water- and air-borne illnesses, respiratory diseases, and injury caused by extreme weather events. The poor are the most vulnerable.[21]

The human implications of climate change are vast and complex. One of the most significant could be the migration of massive numbers of people – potentially tens or even hundreds of millions – either within their states or across international borders.[22] This migration may be sparked by particular environmental events, such as floods or cyclones, by local environments becoming uninhabitable, or by social conflict caused by new scarcities resulting from climate change.

•

This book is about people who might be displaced as a result of climate change. Some say that Tuvalu, a nation of 11,000 people, will be the first to be forced to evacuate its population. Even if the country does not "sink" it could become increasingly inhospitable due to global warming. It is possible that the plants on which many Tuvaluans depend for food could die as a result of salt-water intrusion or changes in the weather. The coral reefs could be devastated by rising sea temperatures and ocean acidification, making the fish that are a staple for many Tuvaluans harder to catch. Added to food insecurity could be ongoing water shortages and changes to weather, subjecting the country to increasingly frequent extreme weather events.

Climate change–related displacement is a complex and challenging subject, not least because the full impacts of climate change cannot be known. This is partly because of our limited knowledge of the complex systems that affect the climate and our limited understanding of the ways in which climate can affect ecosystems. The way humans respond to such changes – if, when and to where they might choose to migrate, for example – is not obvious either. Nor is it certain how the international community will respond to climate change – whether we will take the threat seriously and whether we are prepared to take the necessary steps to curb our greenhouse gas emissions. If we respond quickly and effectively, some of the more dramatic impacts of climate change may be prevented. If we fail, it is possible that we will be living in the future with dramatic, perhaps catastrophic, changes in the environment.

Whatever the case, it is likely that many human beings, particularly in countries that are too poor to fund radical adaptation strategies or are unable to govern themselves successfully in the face of new scarcity and social upheaval – countries that in the development jargon of the day have little "adaptive capacity" – will be forced to leave their homes because of climate-related phenomena. The movement of potentially many millions of people presents real challenges to Australia and the international community. How will the world and individual states respond to the movement of such vast numbers of people? Will we do so by erecting bigger walls to keep such people out? Or will we find more creative means of responding? Do people forcibly displaced due to climate-related factors have rights that are realisable? And if so, which state or non-state entity ought to guarantee these rights? Do such people have a need for protection in a similar way that people who flee persecution – refugees – have a need for and a right to protection? Should a new international agreement be drawn up to set out the obligations of the community of states in response to this new form of forced migration?

Climate change–related migration also raises questions that are beyond the scope of this book. Might the potential end of the state of Tuvalu, one of many small island and other states that could be at risk of disappearing or disintegrating, represent a challenge to the future of the nation-state? Does the possibility of hundreds of millions of people on the move in search of protection, some crossing international borders, force a reconceptualisation of the state as a territorially defined political entity related to a distinct nation? How can national belonging remain meaningful while there is the potential for some states to disappear from the face of the Earth? And if this does happen, who is entitled to the resources that may belong to an extinct state?

These are important issues. But they are also largely, at this stage at least, theoretical. For Tufitu, they are irrelevant, the musings of someone with too much time for things other than the basics. More present for her is her vulnerability. She only has to listen to hear the sea, a source of life and of danger. A short walk and she can see it. It is a time bomb she and others in similar positions around the world must now live with. “For sure I am worried about it,” she told me. “I am worried about it. How can we protect ourselves? How can we do it? Going to another place or staying here. Only two options. Just to stay here and do something for Tuvalu or just going away, migrate to another place. Only two. But if we stay here, we talk today and it might happen tomorrow. We don’t know that.”

This is the story of people like Tufitu and the others I met in mid-2008 when I travelled to Tuvalu to speak with them about the threat of climate-related displacement.

CHAPTER 2

Tuvalunacy or the real thing?

We live in constant fear of the adverse impacts of climate change. For a coral atoll nation, sea level rise and more severe weather events loom as a growing threat to our entire population. The threat is real and serious, and is of no difference to a slow and insidious form of terrorism against us.

— Saufatu Sopoanga, Prime Minister of Tuvalu,
58th Session of the United Nations General Assembly,
New York, 24 September 2003

In short, Tuvalu is a Tuvalu-made ecological disaster that is now an economic disaster. The natives want out because they wrecked the place.

— Patrick Michaels, "Don't Boo-Hoo for Tuvalu,"
www.cato.org, 10 November 2001

From the Filamona Lodge, my temporary home in the middle of Fongafale, I take a brief walk south, passing the AusAID office and the Coconut Wireless internet café. I find the grandly named but not so grand Tuvalu National Library and Archives. It is, as expected, packed with books and magazines. The books have tattered edges and faded spines, and there is a musty smell about the place. I am reminded more of a second hand bookshop than a library. In the Pacific Islands section I find an old copy of a book by Charles Darwin. Its pages have a brown tinge to them and are brittle; the cover is hard and khaki coloured with gold lettering on the spine. I can't find any date of publication so it's hard to tell if it is a very old book or whether it just seems old because it has had to survive in the tropical humidity. The book has an unwieldy title: *On the Structure and Distribution of Coral Reefs; also Geological Observations on the Volcanic Islands and Parts of South America Visited During the Voyage of H.M.S. Beagle.* I sit at an old wooden table on a vinyl-covered chair. An assortment of other chairs, two of them with their backs missing, surround the table.

I am excited by my find because it was here, on Funafuti, that Darwin's theory about the creation of atolls was proven. Darwin had returned to London from his Pacific voyages as the *Beagle*'s naturalist in 1835–36 with a controversial theory. Coral islands, he surmised, were formed from coral growing on top of old volcanoes that had subsided beneath the sea. As the volcanoes slowly sank, dead coral was taken deeper into the ocean while new coral, which can only grow close to the surface of the ocean, grew up. Further volcanic activity then pushed the coral above the surface of the ocean, causing it to protrude in the form of atolls.

Later, I take a walk north, up Tuvalu Road and then right at Palagi (meaning "foreigner") Road. Just around the corner I find a small concrete mound with a piece of poly-pipe sunk in the middle. It is Professor Edgeworth David's drill hole. David travelled from Sydney to Funafuti in 1896, 1897 and 1898, eventually boring deep enough into the atoll to discover volcanic deposits, proving that Darwin's theory on the formation of atolls was correct.[23]

The formation and structure of the atoll has implications for the threat posed to Tuvalu by climate change. In the first instance, the islands of Tuvalu are flat and extremely low lying. At the highest point, Tuvalu is only five metres above mean sea level; its lowest point is said to be at sea level.[24] The average altitude of Tuvalu is, depending on your source, only one[25] or two metres above the sea. It would not take much for the country to be overwhelmed by high tides and waves associated with extreme weather events such as cyclones and hurricanes.

The structure of the atolls means that sea level rise will also have other effects. The coral substructure of atolls is naturally porous, allowing seawater to permeate. When the tide comes in, the water level below the surface rises. This is why, when the king tides hit, seawater percolates from under the ground. It is also the reason why, when the wave surge dislocated her family, Tufitu saw water coming not only from over the coral bank, but also from under the ground. Should the sea level continue to rise, saltwater encroachments on the land will become more frequent.

It is also in the nature of atolls that, if they are over a certain size, they possess a freshwater "lens" that floats above the seawater below. The rain that falls on the islands seeps through the sand and gravel and, because it is lighter than salty water, pools on top of the seawater. The lens naturally rises and falls with the tides under the coral platform beneath the islands. It is the traditional water supply for the inhabitants of Tuvalu. More recently tanks have come into use, but on some islands the freshwater lens remains an important contingency supply of potable water in times of drought.[26] Should the sea level rise too high, it is conceivable that the sea will push the lenses above the surface of the land, so that instead of being protected underground freshwater sources they become a series of fetid tidal puddles.

The atolls of Tuvalu are surrounded on the ocean side by coral reefs. These reefs offer protection from the full force of the ocean, a force that is easy, from the comfort of a large land, to underestimate. While coral has been around for longer than humans, the polyps that are its building blocks cannot survive if the water temperature rises more than 2° Celsius above their optimal levels. Warming seas, including localised temperature rises due to weather changes such as warmer weather without wind, and increasing ocean acidity associated with climate change, could put some coral species at risk. The loss or slowing of the growth of coral could leave the atolls more exposed to the full brunt of the pounding and rising seas.

•

I had arranged to meet Luke Paeniu, the founding director of the Private Sector Organisation of Tuvalu and former director of aid with the Government of Tuvalu. He was late arriving. He had been held up preparing his pulaka pit before the high tide, he told me. Pulaka is a traditional root crop cultivated in Tuvalu. It is grown in flat-bottomed pits dug to the depth of the freshwater lens. Leaves of different plants are added to the pulaka pits, the compost from which fertilises the nutrient-poor soil. From above ground, pulaka plants are huge green leaves on stems of red or green. Think of silverbeet on steroids. Think of *The Day of the Triffids*.

Rising sea levels have begun to affect the pulaka. Luke Paeniu told me that the high tides are now higher than they used to be. He said that in the past, seawater would come onto Fongafale in Funafuti only during the king tides. But now, at every full moon, water penetrates the land. Each month there are two high tides, and the one at the full moon is exceptionally high – higher, according to Luke, than it used to be. This higher sea level means that the salt water gets among the roots of the pulaka.

Luke Paeniu is not the only pulaka grower to have experienced the rising sea levels. Toalipi Lawti is a former prime minister of Tuvalu. Indeed, he was Tuvalu's first prime minister, taking the post in 1976 after Tuvalu had achieved independence not only from Britain, but also from its former colonial partner, present-day Kiribati. Toalipi is now an old man. He walks with a stoop and a stick. But his mind is still active. He speaks slowly with a breathy voice.

As we sit in the restaurant at the Filamona, Toalipi Lawti tells me about his life, a story that intersects with all the important events in Tuvaluan history over the past eighty years. He also tells me that he has had to change the way he grows pulaka. In the past, he would dig about thirty centimetres around the plant and then fertilise it using green leaves he had collected. But the bottom half of the edible roots of his pulaka have been rotting. This, Toalipi said, is because the seawater is now touching the pulaka roots. Now he has to raise his pulaka for it to remain healthy. "I knew that the sea water gets up here so I had to go higher," he told me. "So instead of digging down a foot I had to go just nine inches down or less. And I found that an improvement." The man who grows pulaka in the pit next to Toalipi's had the same problem. So he raised his pulaka plants as well. "It's a real thing," Toalipi told me when I asked about the rising sea level. Others, however, reckon that the effects of climate change on Tuvalu, including a rising sea level, are a myth.

•

Expatriates fly in and out of Tuvalu. Some stay for months or years. Most stay only for days. Many, it seems, simply make the short journey from the government-owned hotel to the neighbouring government offices, built with aid from Taiwan, and then back to the hotel. They are trainers and consultants and assessors from institutions great and small. They are people who are here to help. Then there are are researchers and academics – like me – who spend their hours milking Tuvaluans for information. All of them seem to have opinions about Tuvalu.

It's a competitive business being an expat. It means that you must know a lot about a lot of things, one of which, when you are in Tuvalu, is Tuvalu. So there are bankers who are also, apparently, experts in the formation of atolls. There are seamen who are experts on taxation and fly-in lawyers who can tell you about the cultural intricacies of Tuvaluans. There are scientist-activists who know about refugee law, and film-makers turned philanthropists who, somewhat ironically, want to act as censors to ensure that what is written is consistent with their particular ideological bent. And everyone has an opinion about climate change. For some, it is a joke. A beat up. "Did you know that the US Peace Corps, so worried about the rising sea level, recommend life-jackets for volunteers who come to Tuvalu?" It is easy to get lost in the heady business of lunch and dinner between the hotel and the Filamona Lodge.

Nor is it only the expats dining at the hotel who doubt claims about Tuvalu's experience of climate change. There are, of course, those who, despite the evidence, still dispute the existence of anthropogenic climate change. These people include the Competitive Enterprise Institute's Christopher C. Horner and the former president of the American Association of State Climatologists, Patrick J. Michaels, now a senior fellow in environmental studies at the Cato Institute – a think-tank associated with big greenhouse gas polluters.[27] They argue that rather than climate change damaging

Tuvalu, locals are responsible for their own environmental destruction.[28] But, while not specifically dealing with Tuvalu, the Intergovernmental Panel on Climate Change has taken a more sophisticated and nuanced approach to understanding the effects of climate change on small islands.

"Climate change and sea-level rises are not unique contributors to the extreme vulnerability of small islands," write the Panel's experts on small islands. "Other factors include socioeconomic conditions, natural resource and space limitations, and the impacts of natural hazards such as tsunami and storms. In the Pacific, vulnerability is also a function of internal and external political and economic processes which affect forms of social and economic organisation that are different from those practiced traditionally, as well as attempts to impose models of adaptation that have been developed for Western economies without sufficient thought as to their applicability to traditional island settings."[29]

John Connell, a geographer at Sydney University, has taken a sober yet sceptical position on the impact of climate change on Tuvalu. He has raised questions about the vested interests promoting a particular version of Tuvalu's experience of rising sea levels due to climate change. Writing in 2003, before the latest Intergovernmental Panel on Climate Change report, Connell accepted that a sea level rise due to climate change represents a real threat to low-lying Pacific island states like Tuvalu.[30] But he casts doubt on whether the lived experience, as told by activists, sections of the media and the government of Tuvalu, is accurate. It is in the interests of each of these groups that a dramatic, doomsday story be told. For activists, the sinking nation highlights the need for changes in environmental policy and practice. It highlights the need to tackle the problem of global warming in particular, but also more generally for the world to take the environment more seriously. The media's interest is in the "scoop" of disappearing islands, the story that will sell more newspapers, more television time, more advertising.

The Tuvaluan government also has an interest. It can use climate change as a means of financing development, leveraging aid and influencing migration outcomes by emphasising Tuvalu's victimhood. At the same time it can shirk responsibility for the other, admittedly difficult, development tasks it currently faces, such as promoting economic development for the small, remote and relatively poor nation with few options for economic growth, and dealing with ongoing urban migration from the outlying islands to Funafuti.

Connell's paper is a sober one. It may now be dated as new scientific evidence challenges some of his assumptions. But it is a reminder of the need "to transform those policies in metropolitan states that continue to contribute to global warming and to develop appropriate environmental management policies within atoll states."[31] It is also a reminder of the political and ideological interests that permeate the climate change debate. On one side are the "denialists," those individuals and groups seeking to downplay the effects of climate change or the role of human activities in causing it. The denialists – whose intellectual centre is in United States–based petrochemical companies like Exxon-Mobil and the think-tanks they fund – have sought to cast doubt, in a strategy borrowed from the tobacco industry, on the scientific evidence of climate change and to discredit those advocating serious responses to the threat.[32]

The Australian variety of denialists, who draw heavily on their American mentors, include groups such as the Lavoisier Group, the Centre for Independent Studies, the Institute for Public Affairs (IPA), business moguls Hugh Morgan and Ray Evans, and commentators like the IPA's John Roskam, the *Australian*'s Alan Wood, Miranda Devine from the *Sydney Morning Herald*, the *Herald Sun*'s Andrew Bolt, and Michael Duffy from ABC Radio National's *Counterpoint*. The denialists seem to be motivated by a combination of economic interests and an ideology that views environmental concerns as not only opposed to economic growth but part of a search for meaning on the part of the those on the political left.

But it is not only the denialists, with their vast economic and political resources, who have obscured the reality of climate change and related factors. Environmental activists are not immune to exaggeration and distortion. The former US vice president, Al Gore, won an Academy Award and a Nobel Peace Prize for *An Inconvenient Truth*, an important film that brought to a global audience the threats posed by climate change. In one scene in *An Inconvenient Truth* we see images of melting land-based Antarctic ice. Gore tells us that, unlike sea-ice, when land-based ice melts it raises sea levels in the same way that a melting ice-block floating above the water-line in a glass raises the water level. As the narration continues Gore tells his viewers that this is the reason that "the citizens of these Pacific nations have all had to evacuate to New Zealand." The image then shifts to an unspecified tropical island where the sea is inundating the land. We are not told what is happening. Is it a king tide? Is it a wave surge caused by a cyclone? Is it rising sea levels due to climate change?

There is no evidence that Pacific islanders are "evacuating" to New Zealand. Nor has there been a mass exodus from low-lying Pacific islands due to climate change. There is, and always has been, a lot of movement around the Pacific, including to New Zealand. Indeed, more Cook Islanders, Niueans and Tokelauans live in New Zealand than in their ancestral islands.[33] But this migration is for economic, family and educational reasons. It has not been caused by climate change.

The denialists have also published the misconception that New Zealand is accepting people displaced by climate change. According to the Cato Institute's Patrick Michaels, "all 10,991 poor inhabitants of Tuvalu, an island [*sic*] in the middle of the Pacific Ocean… have pestered the New Zealand government into accepting each and every one of them as environmental refugees… The New Zealand government took the bait. The first evacuees are scheduled to arrive next year [that is, 2002]."[34]

Similarly, an Australian Greens senator, Kerry Nettle, says that "New Zealand is effectively taking 'climate refugees' under its Pacific Access Category immigration program."[35] The category, established in 2001, replaced existing labour schemes and does not cover climate change. Each year 250 Tongans, seventy-five Tuvaluans and fifty people from Kiribati are chosen by ballot to come to New Zealand and apply for residence.[36]

Even at a broader political and economic level, the Pacific Access Category is not linked to climate change. Enele Sopoaga, Tuvalu's

Secretary of the Department of Foreign Affairs, told me that it is part of a general response to the economic development of the Pacific region. A number of the nations in the Pacific, including Australia and New Zealand, are promoting a free trade agreement, the Pacific Island Countries Trade Agreement. Although it recently ratified the agreement,[37] Tuvalu had been reluctant to sign, saying that it would be disadvantaged unfairly; it has no real exports that will benefit from a free trade agreement, and in the event that the agreement takes off it will lose income it generates from the tariffs and duties it currently collects from imports. According to Enele Sopoaga, New Zealand's decision to accept more Tuvaluan migrants is partly an attempt to address the losses Tuvalu would incur from any free trade agreement. He is keen for Australia to follow New Zealand's lead and would be disappointed that Tuvaluans were excluded from the Pacific Islands temporary migration scheme announced by the Australian government in August 2008.[38]

The Gore film is not the only example of a convenient overstatement designed, it would seem, for dramatic effect. In the course of researching this book, I came across other misinformation. I read, for example, that 3000 Tuvaluans have been dislocated already due to climate change. This figure is regurgitated by media across the world and on the internet by environmental groups and think-tanks.[39] Three thousand people would be equivalent to almost a third of the population of Tuvalu and about three quarters of the population in the capital, Funafuti. This exodus has not occurred.

As far as I can tell, the claim has its roots in obfuscation and has been perpetuated by sloppiness. It rests on the conflation of two unrelated facts: one, that about 3000 Tuvaluans have emigrated, and, two, that because Tuvalu is particularly vulnerable to climate change, Tuvaluans face the prospect of displacement in the future. It is difficult to know who first conflated the two facts. But Friends of the Earth's *A Citizen's Guide to Climate Refugees* is a good example. *A Citizen's Guide* quotes a former Tuvaluan prime minister saying that the effects of climate change are already dramatic and observable in Tuvalu. It then says, "Nearly 3000 Tuvaluans have already left their homelands."[40] Both statements are true. About

2500 Tuvaluans live in New Zealand[41] and up to 500 are seafarers on international ships at any one time.[42] But the wording of the publication is intended, it would seem, to make a link between climate change and migration, a link that, until now, does not exist in Tuvalu. It is then the inattentiveness of others – such as the world's media or other environmental groups[43] – who, without a careful enough reading of the two sentences, blur them. The sloppiness is reinforced when journalists re-hash the claims without further investigation or consideration.

There are also stories that Tuvalu has already gone under and that vast numbers of Tuvaluans have already had to flee.[44] The misinformation about the fate of Tuvalu and Tuvaluans has become so bad that Christopher C. Horner – an observer who is far from independent given his link to the Competitive Enterprise Institute,[45] an organisation funded by some of the US's big greenhouse gas polluter-denialists[46] – has labelled the phenomenon "Tuvalunacy."

Of course, it is not as simple as that either.

•

I came across Tasi Pitoi at his house on the southern part of Fongafale. Tasi is one of Tuvalu's 2000 trained seafarers who, by working for international shipping companies, generate between $2 million and $4 million annually for Tuvalu.[47] The labour of these men is one of Tuvalu's only export commodities. As of January 2006, the equivalent of 15 per cent of the country's adult population was on-ship.

Tasi no longer does the international rounds. Instead of spending nine months of the year at sea he works on local boats. When I met him, the thirty-nine-year-old was bare-chested and glistening with sweat. He was engaging in some heavy work. He took a break to talk to me, putting on an old yellow t-shirt which soon became a patchwork of dark and light, wet and dry.

Tasi had set out the form-work for a sea wall he was building where his family's land reaches the lagoon. He was filling this rectangular construction, jutting out from the shoreline, with coral stones. I asked him why. He said that during the westerly season waves from the lagoon come up onto the land. There is a need to protect the land. "By experience," he said, "it is getting worse."

It is a big project and is being undertaken at significant personal cost. "It is very, very expensive, but we cannot run away from this," Tasi said. "We have to protect ourselves for the future. We have to build this because we have to protect not only me, but my kids, my family, my generation, who want to stay in our beautiful land. So we have to do this, even though we don't have much cash to afford this. But we can't run away… In my opinion, it is real, it is happening. I've been born here, grown up here and it's really happening."

Semese Alefaio would agree. He is a renowned spear fisherman, having practised since he was only four years old. He is also an expert free-diver, an activity that is not without its risks. One man told me that to be a free-diver you have to be prepared to bleed from your ears due to pressure of the water at depth. Occasionally young men die diving like this, overestimating their ability to hold their breath. When I think about Semese, I am reminded of the main character, Jacques, in the French film, *Le Grand Bleu* (*The Big Blue*), a man who becomes more and more at home in the sea, a man who lived and dreamt and, ultimately, died with the sea.

Semese spends a lot of his time in and on the water. He has also spent a lot of time listening to his elders, to his father and his uncles, learning how the water, and what is in the water, behaves. At the age of just thirty, he has an understanding of the sea and its surroundings that inspires awe from others in the community. He puts that knowledge to work for his community, as the coastal management project officer at the Tuvalu Association of Non-Government Organisations.

I hitched a ride with Semese, Sandrine Job, a French marine biologist, and Zaidy Khan, a Fijian coastal management and community development expert, to the Marine Conservation Area, an exclusion zone that Semese helped to establish to preserve the marine flora and fauna.

We boarded a wooden boat and headed west across the lagoon. We had not gone far before the wind kicked up and the rain set in, and soon it was impossible to see more than a hundred metres in any direction. Semese navigated, it seemed, by magic or intuition. Later he told me that he knew where he was going by watching the direction of the wind and waves.

We stopped briefly at the islet of Funafala, where a small community of twenty people lives. It was Sunday and locals were worshipping in the local *maneapa* so we did not speak with any of them. Semese said that an old man who has since died told him that in times past the *maneapa* was in the centre of the village. Now, the traditionally constructed, thatched-roof village hall is at one end and much closer to the coast. In the past, there were six rows of houses between the *maneapa* and the shore, but over several decades, about fifty metres of shoreline has been claimed, apparently by the rising sea, leaving no other structure between the *maneapa* and the water.

Later, when I was back on Fongafale, I visited Mitala Foulau. Wearing a bright yellow cotton blouse with green trimming, she sat on a woven mat on the floor. She was making cigarettes to sell for twenty cents a piece, rolling tobacco into the soft part of the pandanus leaf that her friend Melete Tulua had separated from the tough outer skin. When Melete spoke, she was subdued. Mitala, however, was animated, gesticulating with her hands. She confirmed the story that Semese had told me about rising sea levels and the erosion of Funafala. "I have also observed that it's true that the sea is rising," fifty-one year old Mitala said. "At high tides, at exceptional high tides, king tides, the waves actually reach the *maneapa*, the big meeting hall. Unlike in the past no such waves can reach there."

"The other observation is the distance between the islets. The nearest islet to Funafala, in the old days, in the past, the distance was close. They were near. But now because of sea damage, the sea erosion, there is a further distance between the two islets." There was a very practical way that Mitala could measure this. The people who live on Funafala keep their pigs on the adjacent islet, Telele. "In the past," she explained, "if someone, maybe a woman or man, went to feed the pigs but left something on Funafala, that person can just yell back. Nowadays, if that person yells, no one can hear them because it's too far. So from those observations, the sea is rising. The sea actually erodes the beach. You notice that in Funafala where the houses were before, we have planted the pandanus trees further to the beach. But now the sea is right down past that pandanus and is very close to the houses now."

•

Semese Alefaio points to other areas of coastal erosion to demonstrate the impact of sea level rise due to global warming. In 1999, working with the Funafuti *kaupule* or town council on the Marine Conservation Area, Semese was involved in measuring the beach. The project found that the shoreline of some islets was declining by five metres every five months. Sand was washed away leaving just the reef rock exposed. Once, this shoreline was a nesting site for turtles, but now, nearly a decade later, the turtles have gone because there is no nesting habitat.

This is not to suggest that sea level rise is the only and uncontested factor in coastal erosion. Much has been made of the "lost islet" of Tepuka Savilivili, to the west of the Funafuti lagoon. Where coconut trees once stood, the islet is now barren, salt water and coastal erosion apparently having wiped out anything living on the tiny spot of land.

There are also suggestions, rumours really, that the sand from Tepuka Savilivili has migrated to the other side of the lagoon and a new islet has formed. Sydney University's John Connell has suggested that the denuding of Tepuka Savilivili is "typical of many tiny sand islands within atolls that constantly change as they have done for centuries in response to various local influences." According to Connell, way back when Professor Edgeworth David was testing Darwin's theory on atoll formation, there were just three coconut trees on Tepuka Savilivili, compared with thirteen in 1976.[48]

But there are other factors that Semese and others cite as evidence of the impact of global warming. Fishermen across the nation have reported falling fish stocks. Semese says that bird life – plovers, some species of boobies, noddies, frigates and Pacific pigeon – is declining. On some islands plants are reported to be bearing less fruit. And the weather patterns are changing, with longer periods of drought and less predictable seasons.

Another important impact of climate change is on the coral, the central element of life in Tuvalu. According to Semese, "No coral, no fish, no life." When we went to the conservation area, the coral, according to Zaidy and Sandrine, was healthy. But elsewhere it has died.

In fact, Sandrine and Zaidy were in Tuvalu to work with Semese and the Tuvalu Association of Non-Government Organisations on a coral planting project, which took coral from a donor site and planted it in an area where there was no coral. Sandrine and Zaidy were in Tuvalu for the fifteen-month inspection. What they discovered disappointed them bitterly. The coral they had planted had suffered a massive mortality rate. The donor region had also suffered. There was evidence of coral bleaching, which occurs in response to abnormally high sea surface temperatures – that is, about 1°C above average usual maximums. Some corals can recover from such an event, but if bleaching occurs for a more extended period or if sea surface temperatures exceed 2°C above average maximums, corals die.[49] Sea surface temperatures can be affected by local factors. Hot weather over several days with no wind can lead to significant rises in sea surface temperatures. Locals were describing just such weather in the weeks before we were in Tuvalu.

But according to Sandrine and Zaidy other factors (including the nutrients entering the ecosystem as a result of poor waste disposal) may well have killed the coral. The septic tanks, for example, are thoroughly inadequate. Some, I was told, are built with four sides and no bottom, allowing the sewage to leak through the porous soil into the water table and mix with the freshwater lens and the seawater that ends up in the lagoon. The nutrient runoff from the sewage and other waste, including the runoff from piggeries, might have led to an algae outbreak, according to Sandrine and Zaidy. The coral they had inspected was not only bleached; it was also covered in algae.

It is likely that the dead and dying coral that Semese, Sandrine and Zaidy were inspecting were the victims of several factors, some linked to climate change and others to poor waste disposal. The most confronting example of this locally induced environmental damage is the "borrow pits." It was on one of these that I met Solonaima Faimaua.

Solonaima's home has the bizarre quality of being located, simultaneously, in a tropical island paradise and in the middle of a rubbish tip. The family – Solonaima, her husband, their two young children, Tekaile Jr, 3, and Isaaco Jr, 2, and her mother – lives in a house on stilts atop the borrow pit to the south of Fongafale. The borrow pits, so the story goes, were dug by the US air force during the second world war. The Americans used Tuvalu, then known as the Ellice Islands and part of the Gilbert and Ellice Islands British colony, as an advance base. They "borrowed" the soil from sites around Fonafale to level the aircraft runway. The borrowed land has never been returned, so the pits have filled with seawater that rises and falls with the tides. Some people remember these pits as saltwater lakes in which they swam and around which they picnicked.

Nowadays, the borrow pits are not so pristine. As Tuvalu has moved further and further into the cash economy it has imported increasing volumes of manufactured products. And with this rise in imports has come an upsurge in waste. Previously, the nation's waste was largely biodegradable, created from local resources such as banana and coconut plants. Now large amounts of non-biodegradable waste is simply dumped in the borrow pits, an unsightly and hazardous practice.

The borrow pit near where nineteen-year-old Solonaima and her family live is a sea of litter. You cannot see the water; instead, there is a colourful layer of plastic containers and shrink wrapping and bottles and cans and bags. To the south of the pit, a red shipping container has reached its life's end. Rubbish spills beyond the water level too. And the place has a rancid smell.

In the midst is a large, rusting tank that was thrown inland by the wave surge that displaced Tufitu's family last year. Solonaima and her family were asleep when the waves came. They were woken and rushed to the highest part of the land, where the road is. "There were big waves," Solonaima said. "We were afraid. We were concerned because of the young kids. For us, we can find shelter, where to get to safe ground. But we were worried about the small children." Solonaima's mother, Melate Tulua, told me that when the wave surge hit, rubbish from the borrow pit was spread all over.

I sat on the floor with Solonaima. Hers is a humble house with almost no furniture. A cupboard stands in one corner, shelving in another. Two old pots sit on top of a small stove. The walls are covered with grime. Some of the windows are unglazed; others louvred.

Solonaima lives atop a health risk. Tekaile Jr's legs are covered in scabs and scars. According to Solonaima, he and his younger sibling had caught scabies from swimming in the borrow pit. Solonaima has now banned the children from swimming there. The damage to her children's health from her home environment obviously plays on her conscience. But she feels she has no choice.

While we were speaking, Tekaile Jr fell asleep in her arms. Solonaima gently placed him on the floor while she got up, prepared a more comfortable bed, and then laid him down.

•

The closest "outer island" to Funafuti is Vaitupu. As we approached it by boat, two breakwaters embraced us and guided us into the village. Dominating the skyline is Vaitupu's church, as impressive as any building that you might find in the nation's capital. It is a stone and timber construction with belltowers to the left and right and archways, all painted white. With a coral-dust sports field in the foreground, the scene is all whiteness and lines and shadows. Behind and to the left of the church are houses, made mostly of white-washed masonry, overlooking pale dirt roads. I had the feeling that the township might as easily have been an idealised representation from a Salvadoran folk-art cross as a real-life place in the remote Pacific.

Free of the developments that mark Funafuti's environment, Vaitupu seemed like a good place to talk with people about climate change and the possibility of displacement without the complicating factor of the local human impact. Vaitupu is different from Funafuti atoll in other ways, too. Its lagoon is almost entirely surrounded by land – in fact, there are two lagoons, a smaller one to the very north of the atoll, and a larger one in the centre. Both have inlets through which water flows with the tides, refreshing the lagoons.

During the king tide of 2006, water spilled over the seawall at the edge of Vaitupu's lagoon. Tui Taumafai, a community development worker for the Vaitupu *kaupule*, showed me photos of the water spreading across the land, flooding roads and buildings. "People were worried so much because it was the first time for people to see something like this happening," he said. He believes that people now see that Tuvalu "is going to be submerged in some years to come. So now they are trying to plan where to go, where to migrate to some places where it is much safer for the families."

Tui has lived on Vaitupu all his life. Sometimes he ventures to Funafuti to meet with officials there, but mostly he stays on his home island. While others of his family have migrated, he seems to have fallen into a life that entirely suits him. He took me by motorbike to his pulaka and taro pit. As we stood looking he told me that his plants are no longer as healthy as they used to be, and nor are his neighbours' plants. Water was lying around in the pits – a mixture of seawater and fresh water, according to Tui. Like Luke Paeniu and Toalipi Lawti in Funafuti, he has raised the bottom of his pits, digging soil from the outside and heaping it into the beds in order to raise them above the height of the water table.

Tui also took me to places where he believes the coastline has been eroded by big waves and high tides. Coconut and pandanus trees lie prone, the sand at their roots washed away. Sharp curves in the coastline have been straightened. Twenty or thirty metres of land, in some places, have disappeared. In other places the sand has shifted. "So we don't know if it is the natural movement of the islands, or if it is caused by climate change," he said. But he also says that over the past five or six years the pace of erosion has accelerated. "For the last five years you can see that it is a problem," he said.

•

Tuu Teo also told me that there have been changes to the environment consistent with climate change. Tuu is seventy-three years old. He has lived in his home on south-eastern Vaitupu, not more than forty metres from the sea, for almost thirty-five years. He lives there now with his wife, two of their four children and ten of his grandchildren. He is the patriarch of a small tribe. Since the 1990s, Tuu has noticed that the pulaka in his pits rots more than it did before.

The pits are constantly wetter than they once were, and the water in them has a salty taste to it. He also said that he has noticed changes in the weather patterns: now when it rains it does so more heavily than before, and droughts are more frequent as well.

Tuu remembers when the wave surge came last year. He recalls that it struck early in the morning, when the rest of his household was asleep. Water flooded all across the road into his pulaka pits. His pig pens, which are closer to the coastline, were damaged. "It was a dangerous time," he said. "We were lucky to have escaped." Tuu is remarkably pragmatic about Tuvalu's future and the possibility that he will have to leave. "If the time comes, I will leave," he told me. "But you cannot predict that time."

"What about your culture?" I asked, thinking that this would be a complicating factor for someone who has lived off the same land and the water that his ancestors lived off. "It is a better idea to run for my life than worry about my culture or my piece of land," Tuu said.

When I asked him if I could take his photograph, Tuu, sitting cross-legged and bare-chested on the raised floor of an outside shelter, called for a shirt. A pale blue Hawaiian shirt with orange and red flowers was promptly delivered to the old man. Two of his granddaughters, one of whom had teeth that were rotted down to their roots, did up the buttons for him. He then coaxed the children into posing in the photo with him. When I showed them the photos I had taken, Tuu laughed heartily. He told his granddaughters that I would take the photos to England to show the Queen.

As I left, a mob of free-ranging pigs veered out of my path, seemingly afraid of a swift kick from those who, if this were an Orwell novel, they would be justified in plotting to overthrow. I saw a small, fat kid who was on the ferry when I came over to Vaitupu. I had noticed him on the boat because even before the ship had left the dock, even before the six-hour journey across the Pacific, he had vomited all over where his family had set up their mats. Family members scraped what they could over the side of the boat and into the sea, and eager fish the length of an adult's forearm devoured what they could. The last I had seen of the boy, he was lying next to someone who I assumed was his grandmother, feeling sorry for himself. Now here he was, and it turns out he is part of Tuu's extensive family. It is a small place, Tuvalu.

•

Back on Funafuti, I went for a walk with Luke Paeniu. We walked north from the Filamona, past the bank and along the road adjacent to the airstrip. We walked past the remnant of what was once a much larger group of pulaka pits, past the paw-paw and breadfruit trees and the banana plants. Luke showed me where pigsties leak waste into the water table. We looked at the failed AusAID project aimed at converting pig-waste into biogas, and at the more successful Australian-funded mulching project. We walked past one of the local nightclubs, which is mostly open to the elements and surrounded by a cyclone fence. When I went inside later in the week, I felt like I was caged on the set of a *Mad Max* movie.

Luke and I walked past one of the borrow pits and watched the water rising as the tide came in. He showed me the lower parts of Fongafale and places where the elevation is higher. We stood and watched a young man climb a ladder to prepare a coconut tree to collect "toddy," a juice that is known as "sweet toddy" when it is drunk fresh and "sour toddy" after it has been allowed to ferment. Nearby, we noticed a trickle of water emerging from under two houses. We probably would have walked past it had we not been watching the man cutting toddy. I asked one of the residents if the water was fresh or seawater. She said it was seawater – seawater oozing up through the ground. It was high tide: not king tide, merely high tide. Luke told me that in the past it was only during king tides that the seawater came up through the land. And Luke knew about this place. He pointed to the house just up the road that he lived in as a teenager.

There are those who would dispute this local knowledge of the effects of climate change, including the rising sea level, because it is not scientifically tested. This is not to say that there is no scientific evidence of climate change on Tuvalu. A sea level monitoring station has been based there for decades. But the results from this station were flawed, essentially because the yardstick it used was moving. The measure was attached to the land, so it recorded sea level relative to the land, meaning that if the land rose or subsided, the height of the measured sea level would be affected. So in 1993 the Australian government installed a high-tech monitoring system called SEAFRAME in Funafuti that measures not only sea levels in two different ways, but also air and water temperatures, wind speed and direction, and atmospheric pressure. The new system incorporated a means of measuring the movement of the land and of taking account of this movement in the assessment of sea level rise. In 2001, a further layer of sophistication was added; a Continuous Global Positioning System station was installed to measure the rate at which the land moves in relation to the centre of the Earth.[50]

A report summarising the findings of the new system was published in June 2006. Although its authors cautioned against giving too much weight to the findings because the data has been collected for a relatively short time, they are nonetheless instructive and potentially alarming. When all other factors are removed, the sea level rise at Tuvalu averages 5.7 mm per year. This is well above the Intergovernmental Panel on Climate Change's estimate of a global average sea level rise over the last hundred years of between one and two millimetres per year.[51] It is also well above the 3.3 mm global average sea level rise between 1993 and 2006.[52]

According to Tauala Katea at the Meteorological Office in Funafuti, this early SEAFRAME data is probably an understatement. "I think when we have more years of data, it may increase up to seven or eight millimetres, according to what we observe," Tauala said. "Since the year 2000, things have changed a lot, including the weather pattern itself... And if I see it through what we observe during our king tide and compare it with all those years and nowadays, and comments from people like fishermen and elders who usually go and work on their pulaka pits, they all noticed that around the late '90s things have changed a lot."

These changes, which are broadly consistent with what the Intergovernmental Panel on Climate Change has reported on a global scale, could contribute in the not too distant future to the outward migration of Tuvaluans from their homeland.

CHAPTER 3

The rainbow covenant

And God spake unto Noah, and to his sons with him, saying, "And I, behold, I establish my covenant with you, and with your seed after you; and with every living creature that is with you, of the fowl, of the cattle, and of every beast of the earth with you; from all that go out of the ark, to every beast of the earth. And I will establish my covenant with you; neither shall all flesh be cut off any more by the waters of a flood; neither shall there any more be a flood to destroy the earth."

And God said, "This is the token of the covenant which I make between me and you and every living creature that is with you, for perpetual generations: I do set my bow in the cloud, and it shall be for a token of a covenant between me and the earth. And it shall come to pass, when I bring a cloud over the earth, that the bow shall be seen in the cloud: and I will remember my covenant, which is between me and you and every living creature of all flesh; and the waters shall no more become a flood to destroy all flesh. And the bow shall be in the cloud; and I will look upon it, that I may remember the everlasting covenant between God and every living creature of all flesh that is upon the earth."

And God said to Noah, "This is the token of the covenant, which I have established between me and all flesh that is upon the earth."

– GENESIS, CHAPTER 9, VERSES 8–17

Tuvalu is overwhelmingly and deeply Christian, a legacy of the the missionaries who began arriving in 1861. Back then, the new religion spread quickly, radically changing traditional social structures. Power transferred from chiefs to pastors and other church leaders.[53] Today, Christianity has no less an influence on

Tuvaluan life, a fact that is evident in Tuvaluan's attitudes towards climate change.

Take Toma Liveti, for example. He lives on the "lagoon side" of Fongafale, a little to the north of the southern borrow pit where Solonaima lives. Toma's place is a stereotypical tropical paradise. It fronts onto the lagoon, the turquoise water lapping at the shore. Trees provide shade and fruit, and the dwelling has more land around it than many other properties in Tuvalu's capital. It is set back slightly from the road, so the noise from the passing traffic, not that it is great, is dulled. It is a million dollar location. Or it would be if it was not in one of the remotest places on Earth.

I met Toma as he was riding his bicycle back from a fishing trip. He was carrying half a fish, its pink flesh facing towards the ground. He took me to his *fale*, a small, open-sided shelter covered in pandanus leaves. His wife was sitting there, but left as we arrived. He spoke quietly, in a deep, breezy voice, a few days' stubble on his face. There was a sense of calm about Toma. I wondered how many long hours he had spent on his own, maybe in a canoe, with the water under and around him. I wondered how much he had given in his life to the long contemplation that can come with extended spells doing simple, life-giving work. As I sat with him, I thought less about Noah and the Judeo-Christian tradition and more about Hermann Hesse's *Siddhartha* – how, after a journey through all the trappings of life, Siddhartha finally found the river and the spiritual peace that flowed with it.

Toma was born in Funafuti. After going to Ocean Island in Kiribati, he returned to Funafuti at the age of ten in 1969. He has been there ever since. Living a semi-subsistence lifestyle, he has watched his environment closely over the years, observing changes in the weather patterns, rougher seas, dying pulaka, higher tides and erosion. But Toma is not particularly worried about the impact of climate change. He has faith. "God promised Noah that no more flood will come," Toma told me, a partly smoked pandanus cigarette of the sort made by Mitala Foulau and Melete Tulua in the crease between the top of his ear and his scalp. "There is only one flood – during that time, the time of Noah."

Toma's theological perspective is widely shared, it seems, across Tuvalu. "Not all Tuvaluans are convinced that Tuvalu will go under water," Reverend Tafue Lusama told me, squeezing in an interview before he flew to New York to attend meetings about climate change. Tafue is a pastor with the Christian Church of Tuvalu, the dominant denomination. "The main obstacle is their belief in the Bible. I can safely say that around 99.9 per cent of the Tuvalu population believe in the Bible and believe in Christ. And they take the word of the Bible to their hearts. So when you ask people whether they believe that Tuvalu will go under water, they will say, 'Definitely no.' And you ask them why and they would say, 'Because God has promised Noah that there won't be any more floods.'"

With evidence growing that Tuvalu is threatened by climate change, a folk-theology that encourages a fatalistic response is inadequate. Certainly, the government of Tuvalu is not placing its faith in the Rainbow Covenant. Rather, it is pursuing real-world solutions to the threat it faces. Tuvalu has become an active participant in international efforts to respond to climate change. It joined the United Nations in 2000 in order to highlight the dangers of climate change and to lobby the international community to do more to curb greenhouse gas emissions. It is involved in the Alliance of Small Island States, which lobbies for cuts to global greenhouse gas emissions, as well as in other regional organisations that contribute to the debate.

Locally, Tuvaluans have engaged in their own adaptation efforts.[54] As well as establishing the marine sanctuary to protect coral and other marine resources, Tuvaluans have begun replanting mangroves and other coastal flora to curb erosion. They are monitoring and documenting changes in their local environment. Local non-government organisations and the government are raising awareness about climate change and its implications for Tuvaluans. The government recently approved the creation of a senior departmental committee to develop a national strategic plan for responding to climate change. It is also pursuing significant adaptation options, such as building "coastal current breakers" as a way of curbing coastal erosion.

Yet any efforts on the part of Tuvalu to adapt to climate change cannot but be limited. It has no land to retreat to if low-lying areas are inundated or the shorelines washed away. It is resource poor, with few options for generating the cash required for the scale of response that is necessary to deal with climate change. Productive capacity is limited by, among other things, its remoteness. With a small, dispersed population, no tertiary education institutions and low per capita incomes, local expertise is constrained. Tuvalu has few commodities to sell to generate international currency: it sells its people's labour, mostly as seafarers on international ships; it sells the right to fish within its territorial waters; and it sells its national sovereignty – in the form of support to Taiwan (by recognising Taiwan as the Republic of China) and Japan (by supporting its stance in the International Whaling Commission) – in exchange for aid and development assistance. In 1998 Tuvalu cashed in on another aspect of its national sovereignty, selling its national internet suffix, .tv, for US$50 million over twelve years.[55] Taking all these things into account, Tuvalu has an annual budget of merely A$33.5 million, a figure that leaves no room to fund multi-million dollar adaptation projects.[56]

Nor is it only Tuvalu's ability to adapt to climate change that is limited. While the effects of global warming are evident and, if the predictions are right, will multiply, the impact of Tuvalu's own greenhouse gas emissions on the global environment are miniscule. The whole south Pacific region, of which Tuvalu is part, is responsible for an estimated 0.06 per cent of global greenhouse gas emissions.[57] In the meantime, the developed world, including Australia, and developing countries like China and India are pumping greenhouse gases into the atmosphere at unprecedented rates. In the face of this reality, any mitigation effort on the part of Tuvalu can only be symbolic.

Similarly, adaptation efforts – whether it is coastal rehabilitation or building a sea wall – might turn out to be little more than "patch-up" jobs. They might help to stave off the impacts of climate change for a period of time, but they could ultimately prove to be futile in the face of rising sea levels, wave surges, and extreme weather events.

In those conditions, living in Tuvalu could well become untenable. According to a position paper published by the Australian Labor Party before the Rudd government was elected, Tuvalu could become uninhabitable within a decade.[58] This claim is far-fetched, written for effect rather than accuracy, but it does seem possible that in the not too distant future Tuvaluans will need to find somewhere else to live.

Some are already considering relocation. At a recent Tuvalu Trust Fund meeting, representatives discussed the possibility of buying land, possibly in Fiji, on which to resettle Tuvaluans. Another suggestion is that they could move to Kioa, a Fijian island already owned by Tuvaluans who moved there from 1947 due to fears that their home island, Vaitupu, would become overcrowded.[59] In 2001, the BBC reported that the Tuvalu government had approached Australia and New Zealand to accept Tuvaluans forced to flee their homeland due to climate change. According to a government official, Paani Laupepa, "New Zealand responded positively in the true Pacific way of helping one's neighbours, Australia on the other hand has slammed the door in our face."[60]

Sitting in his office on the top floor of the tallest building in Tuvalu, the three-storey government offices, the secretary of the Foreign Affairs Department Enele Sopoaga told me that Tuvalu made no such request. Rather, he said, Tuvalu continues to maintain confidence that the international community can stop global warming. "I think the message is this from the IPCC: these are the threats, this is what is going to happen, but if we work together, we can save this by doing this – one, two, three. And the urgency is there and we must do it now." According to Enele, there are real options for mitigating and adapting to climate change: "They are doable, they are reachable."

Talk of migration or displacement due to climate change is premature, Enele told me. Indeed, it is not even on the international agenda. Partly this is strategic. The need right now is to focus on mitigation and adaptation; to speak of migration is to "introduce another item that would divert attention from this." But as I sat in his office in the government building, it was difficult not to wonder if placing so much trust in the ongoing international negotiations to deal with climate change might be more similar to the faith many have in the Rainbow Covenant than Enele Sopoaga might care to admit.

•

The popular perception across the world is that Tuvalu will "go under," that it will sink beneath the sea if the international community fails to curb and then wind back greenhouse gas emissions. If this were to occur, then the entire population will be displaced. They will become, in the unfortunate language of the media and some academics, "climate change refugees." Picture food drops and blue helmets and Hercules airplanes picking up stranded people.

Yet while Tuvalu may ultimately be a victim of global sea level rise, the reality of displacement due to climate-related factors may be far less dramatic. The most recent report by the IPCC has two chapters that are particularly relevant, one dealing with small islands and the other with coastal systems and low-lying areas.[61] According to the IPCC, the risks to countries like Tuvalu are multiple and complex, interacting with other local environmental factors. Climate change threatens the food and freshwater security of coastal and low-lying communities. Sea water incursion into arable land and water supplies, erosion and land loss due to rising seas and surges, drought, and damage to ecosystems (including to coral reefs as a result of warming seas) will all undermine crop growth, subsistence food resources and water supplies. This will contribute to extreme food and water stress.

Increasingly frequent and extreme weather events will result in death and injury. Food and freshwater resources will be damaged. Small island communities will have no inland haven when their coasts are battered.

Climate change is projected to lead to an increase in diseases borne by insects, food and water. Malaria, dengue, diarrhoea, heat stress, skin diseases, acute respiratory infection, asthma and other illnesses are predicted to rise.[62] As well as these physical illnesses, the IPCC predicts an increase in common mental disorders as a result of the new stresses associated with climate change.[63] The health implications of global warming will be exacerbated by the impacts on vital infrastructure, such as hospitals and roads. Inundation, erosion, and other extreme weather events will undermine health and transport systems, affecting food delivery, for example, and access to health care.[64]

All this may well lead to displacement.

•

"I don't want to leave," Solonaima's mother, Melate Tulua, said. "Why should I leave? God has given this place as a home. I have a right to stay here and I enjoy living in Tuvalu. I don't want to leave. I want to stay on my home island, Tuvalu."

This is the sentiment of most Tuvaluans. Some, like Mitala Foulau, say that they are prepared to die rather than leave their homeland. Others, like Tasi Pitoi, are less heroic – or more realistic, depending on your perspective. As he took a break from building his seawall, an obvious effort to prolong remaining in his homeland, Tasi told me that people who say they won't leave are kidding themselves. "People say, 'No, I will stay here.' But the reality is different," he said. "If you throw someone into a deep ocean he will swim. He will say, 'I want to leave.' You have to leave. You have to go. Because if something happens here, what are you going to do? We are being forced to leave. If I am not leaving, maybe my kids will be leaving. It doesn't mean that it will be easy to leave this beautiful land."

The idea of compulsion, of being forced to leave, plays a fundamental role in the discussion about displacement due to climate-related factors. Force involves a lack of choice, a violation of liberty. Forced migration entails a lack of consent, a denial or diminution of agency. This is not to say that forced migrants are completely disempowered; even forced migrants make choices, although their options may be severely limited. The choice to flee one's homeland for fear of persecution is a terrible decision to have to make, but it is a choice nonetheless. Forced migration involves victims, people to whom something has been done. It implies injustice. It calls for something to be done in order to right a wrong. It has moral implications. And it raises obligations. Forced migration is also, in both the causal sense and in the obligations it creates, fundamentally political.

The 1951 Refugee Convention and its 1967 Protocol were a response to the moral and political issue of forced displacement. Drafted after the second world war, the Convention is a recognition that the international community failed to protect those persecuted by Nazism and a means of responding to the millions of people displaced throughout Europe as a result of the war.

The Convention articulates the cornerstone of international refugee protection, namely that people ought not to be returned to situations where they will be subjected to ill-treatment. This is called the principle of *non-refoulement*.[65] The Convention also spells out who ought to be entitled to protection. It defines refugees as people who are outside their countries of nationality or residence and who are unable or unwilling to return to those countries due to a well-founded fear of persecution for reasons of race, religion, nationality, political opinion or membership of a social group.[66] Our understanding of who might be entitled to protection has expanded over the years – through, for example, the notion of "persons of concern" to the United Nations High Commissioner for Refugees (which includes not only refugees but asylum seekers, stateless persons, internally displaced people and returnees) as well as through instruments such as the 1967 Declaration on Territorial Asylum,[67] the Organisation of African Unity's 1969 Convention Governing the Specific Aspects of Refugee Problems in Africa,[68] the Central American Cartagena Declaration on Refugees signed in 1984,[69] the 1984 Convention Against Torture and Other Cruel, Inhuman or Degrading Treatment or Punishment,[70] and the 1998 Guiding Principles on Internal Displacement.[71] But it is the 1951 Convention that remains to this day the foundation of the international refugee protection system.

Notwithstanding this, the Convention and the global protection regime – including the UNHCR, the key international organisation charged with protecting refugees – has been under pressure in recent decades. Governments, including Australia's, have sought to sidestep their internationally prescribed obligations. For example, there have been calls, including from the former Australian government, to reform the UN refugee definition in order to restrict the number of people eligible for protection under the international system.[72] Governments have also attempted, sometimes successfully, to prevent people claiming to be in need of protection from entering their territory in order to have protection claims assessed. Australia's Pacific Solution – with its naval blockade, its legislative manipulation of the border and its offshore processing system – was a particularly dramatic example of this sort of practice.[73]

Despite the threats that the international refugee protection regime continues to face, the Convention and the UNHCR are important guarantors of refugee rights and of the notion of protection from persecution. The established protection regime, flawed as it is, remains a key moral and legal yardstick against which states are held to account for their treatment of refugees. Even those who honour the international protection regime in the breach try to be *seen* to honour it nonetheless. The Howard government was prepared to argue for a restriction of the definition of a refugee and to violate the rights of asylum seekers and refugees in its care, but it was at the same time insistent that all of its actions were consistent with the international protection system.

The protection regime has taken on a moral and political power well beyond the words of the Convention or the miserly budget of the UNHCR. In part, this is because it articulates and makes real a moral sentiment that is compelling: people who are forced to flee their homes due to the misuse of state power ought to be protected. Such people are victims of the dominant institution of political organisation in the world today – the state. Even political theorists who place primacy on states' rights to control their territorial borders argue that states also have obligations to protect non-citizens in need of protection.[74]

The compelling nature of the claim that people who are forced to flee their homes are entitled to protection is part of what is attractive to those who argue in favour of the notion of "climate " or "environmental refugees." The point of attaching these descriptors to the word "refugee" is to indicate that people fleeing environmental degradation are like "refugees" in that they have been forced to flee for their lives and well-being. People fleeing environmental factors are different from "refugees" in that they are not facing persecution and normally do not cross international borders, but they are similar to refugees in that they are "forcibly" displaced.

The notion of force, by shaping the understanding of the extent to which such people are owed protection, influences policy responses to displacement. It is, in part, this sense of obligation that the terms "climate" or "environmental refugee" are supposed to arouse. Indeed, many of those who use terms like these seem to be less concerned with issues of protection and more with the political work that such terms can do. They are impressed that the term "refugee," and the numbers that are potentially at stake, will get political leaders, policy makers and the community more generally to take more seriously the environmental issues that are central to their worldviews. "Environmental refugees" are not of interest in themselves, but as a means to a bigger, greener end.

Yet there are real problems with this terminology. For one, it is legally meaningless. These terms have no place in international or domestic law. They are undefined and highly contested, and have no rights associated with them. They are terms with a loud political bark but with no legal bite.

Conceptually, the linking of "refugees" and those displaced by climate change is problematic. Refugees flee persecution, oftentimes perpetrated by a government or quasi-government agency. They flee their homelands because their government is unable or unwilling to provide them with protection. In this sense, the government itself is the cause of refugee displacement. Refugees are the victims of state power. This is not self-evidently the case for people fleeing climate change–related factors. Indeed, the governments of those displaced by climate change may be benevolent but completely powerless to prevent the effects of the changing climate. Tuvalu is a case in point. If Tuvaluans must leave their homeland, they will do so despite the efforts of their government to protect them from the effects of climate change. So, in one instance, the government is the persecutor and the cause of displacement. In the other, it is the impotent protector against causal factors that are the result of the activities of developed and big developing nations.

Linking "climate" and "refugees" is also confusing to a public generally ignorant of the subtleties of refugee protection and susceptible to manipulation by leaders seeking political gain through promoting popular anti-refugee sentiment. In this sense, "climate refugee" and "environmental refugee" risk becoming disparaging terms, just as the term "economic refugee" (implying that seeking to migrate to improve the lot of one's self and one's family is somehow illegitimate) is used to cast doubt not only on particular asylum seekers or refugees but on all refugees. It is a term that fits within a broader lexicon surrounding refugee and migration issues – including "illegal immigrant" and "queue jumper" – that is at the same time widely used, never clearly defined and laden with meanings. Even if attaching "environmental" or "climate" to "refugee" is not (yet) similarly dismissive, it muddies public discussions about refugee protection.

•

Notwithstanding the confusion it engenders, and recognising that the notion of an "environmental refugee" is legally precarious, some activists and academics have called for environmental factors to be incorporated into the existing Refugee Convention as grounds for protection.[75] Doing so would give the concept of "environmental refugee" teeth. Yet such suggestions are politically naïve. Any attempt to incorporate environmental factors into the existing refugee protection regime would be unlikely to occur without a broader discussion of refugee protection and, given what we know about states' attitudes towards refugee protection, it seems likely that there would be increased pressure for a restrictive rather than liberal conception of protection. Clearly, this would diminish rather than enhance refugee protection.

Millions of refugees and others in refugee-like situations are already without adequate protection. According to the UNHCR, the number of people in need of protection from persecution and from conflict has been on the rise again; last year, there were about 16 million refugees and some 26 million people internally displaced due to conflict throughout the world.[76] Many are exposed to ongoing human rights violations, to hunger and desperately sub-standard housing, and have no real prospect of rebuilding a life in safety and security. Expanding the pool of "refugees" would further threaten existing protection resources. Adequate protection is already spread too thinly.

Yet while protection is a precious and limited resource, it is not finite. On the contrary, it is renewable. But like a forest, it can be over-exploited. It needs to be managed wisely and allowed to replenish. The ground in which protection takes root and grows is the community – both international and local. Ensuring the fertility of this community, maintaining community support for protecting the displaced, is a political activity.

Incorporating environmental factors into the existing Refugee Convention is wrong-headed for strategic and conceptual reasons. But ensuring that people in need of protection for environmental reasons can find protection is an important task in which to engage. There is a need to develop an international legal agreement and accompanying institutional and financial arrangements to ensure that people who are displaced by climate change can be protected.

This task is complicated by the difficulty of disentangling environmental factors from a myriad of others that lead to displacement. According to geographer and refugee studies expert, Richard Black, there is no such thing as an "environmental refugee"; environmental factors may play a part in forced migration, but these are always linked to social and ethnic conflict, weak states or human rights violations.[77] In other words, it is not so much environmental factors that lead to dislocation as the political management of such issues. The focus on environmental factors, according to the former head of Oxford University's Refugee Studies Program, Stephen Castles, summarising Black, "is a distraction from central issues of development inequality and conflict resolution."[78]

The existing refugee protection regime is not immune from this causal complexity. Indeed, in cases involving massive numbers of forced migrants it is impractical to assess each individual against the

Convention definition of a refugee. Instead, large numbers of people, including those who may be fleeing persecution, war and famine, are treated by the UNHCR and other international organisations as "persons of concern" and entitled to protection and assistance.

This causal complexity and the obligations owing to refugees have promoted an increasing interest in disentangling "mixed flows" of migrants over recent years. States have become concerned about non-refugees using refugee protection processes in order to gain access to territory and to other resources such as welfare assistance and health care. Wealthy nations like Australia have spent vast amounts of money in order to separate refugees from non-refugees. Academics have come up with models to try to explain the multifaceted factors leading to people movements, factors that both compel people to leave their homelands (like conflict, human rights violations and grinding poverty) and draw them to other places (like peace, liberty, economic prosperity and family connections). People fleeing persecution may also desire a materially better life.

Yet amongst all this disorder, the Refugee Convention remains solid. In a world of complex and vast needs, the Convention identifies a discrete need and provides a framework for addressing it. The need is protection from persecution. The Convention deals with this need when it cannot be met in a person's home country. The means of addressing it is the institution of asylum. While the Convention provides the yardstick against which to measure whether a person is a refugee, there is no such test for someone fleeing climate-related factors.

•

With projections that tens or even hundreds of millions of people will be displaced due to climate change, the international community ought to establish an agreement that defines who might be entitled to protection as a result of climate-related factors and what that protection might look like. This is not just a question of coming up with a suitable form of words. It involves understanding the need and how it can be be met. Like the 1951 Refugees Convention, this agreement ought to have modest aims. It must respond only to a discrete, defined need and must not be an attempt to resolve injustice or great human need across the globe.

A key challenge for such an agreement would be to articulate some sort of threshold test against which to assess who might be entitled to protection from climate-related displacement. Protection is the ultimate purpose of any international agreement and of any regime developed to respond to climate-induced displacement. But underpinning such a conception of protection is a discussion of inhabitability that in turn relates to the question of force.

While there may be instances in which it seems clear that a place has become uninhabitable – even if only for a short time, as in the case of a once-off natural disaster – the question of habitability is subjective. Humans have adapted to live in all manner of places. People live in the most extreme environments – amid ice and darkness and in the glare of desert heat. Religious beliefs, ontological frameworks and cultures have evolved around the challenges of living in these circumstances. It is also the case that most people do not want to leave their homes, and if they are forced to do so, would prefer to return as soon as they can.

Somehow, protection from climate-related displacement must be founded on an objective means of assessing when a particular place becomes uninhabitable. There are empirical issues at stake here. How do you determine and measure climate-related factors leading to displacement? How will they be separated from other factors, including other environmental factors, which may lead to dislocation? At what point does a physical environment become so degraded that it is no longer capable of sustaining a humane existence?

The answers to questions like these rely not just on some sort of scientific observation. They are value laden. Indeed, they point to a question of an altogether higher order: what is a humane existence?

Of course, frameworks exist that might help answer such a question. The international human rights regime – the United Nations Declaration on Human Rights, the International Convention on Civil and Political Rights and the International Convention on Economic, Social and Cultural Rights, among others – offers a way into the answer to this question because such a discourse is nothing if not an attempt at articulating rights that are inalienable in order to live a life that is fully human. But such a discourse does not self-evidently offer the means to distinguish which people ought to be entitled to protection.

A discussion of what constitutes a humane existence is a real challenge for any process of developing an institutional response to climate-induced displacement. It seems somehow unreasonable to have a discussion about a humane existence in reference to climate change without touching on broader issues of development, poverty and international trade. There are many millions of people throughout the world who have reason to claim that even without climate change they live lives that cannot be characterised as humane or dignified. There is, of course, a difference between such people and those displaced by climate change; the former can, in theory at least, have their rights or needs met *in situ*, while the displaced cannot. But such a point will surely be lost on those living in desperate, soul-destroying destitution yet in circumstances that might be deemed environmentally habitable. The challenge, then, is how to frame a response to the particular, discrete needs of people displaced by climate change–related factors without making such a framework so broad as to capture the many other hundreds of millions or even billions of people in great and similar need, but for whom responses ought to be different.

The question of a humane existence might also be put negatively. When do particular circumstances become inhumane? The Refugee Convention answers: persecution is intolerable! If it cannot be prevented in a person's place of residence, that person ought to be able to find protection from persecution in another state. Environmentally induced displacement does not provide for such a neat solution, partly because what one person might find an intolerable environment, another finds not only capable of physical sustenance but also of spiritual significance. There is a need to come up with a measure of when life in a particular environment becomes untenable.

•

Inhabitability is linked to the question of force both causally and in terms of timing. Climate-related factors will force people to leave their homes temporarily due to natural disasters and/or when they assess their environment to be so degraded as to have become uninhabitable. There is a host of environmental reasons that could lead people to assess their home environments as no longer able to sustain a reasonable life.

Environmental degradation due to climate change may be dramatic. Catastrophic weather events may force people to migrate temporarily or permanently. Amid predictions of extreme weather events of increased frequency and severity, it is possible that even in communities where such events already occur, individuals and communities may conclude that it is not tenable to continue living in their homelands.

Then there are the less-dramatic environmental changes associated with climate change, changes such as increased drought, salinity or erosion that, over time, place growing stresses on a community's water and food security. These changes may be cumulative, raising the question: at what point is a person or group of people forced to leave their homes and potentially entitled to protection due to climate-related factors?

Academics have posited a range of models to understand better where compulsion fits into migration decisions due to environmental change. Indeed, the question of environmental factors in forced migration was discussed in the academic literature well before climate change was so prominent in public debates.[79] One useful attempt at understanding environment as a factor in forced migration is Diane Bates's continuum, on which involuntary migration sits at one extreme, voluntary at the other, and compelled migration in between.[80] Migrants who sit closer to the involuntary end, she calls, problematically, "environmental refugees." Those in the middle are "environmental emigrants." And those at the voluntary end are simply "migrants." Bates further classifies "environmental refugees" into three groups. "Disaster refugees" are those fleeing acute events, whether they are natural or technological. "Expropriation refugees" are fleeing severe environmental disruption caused by human activities that are intended to dislocate, including warfare or economic developments such as dam construction. "Deterioration refugees" are people dislocated due to gradual anthropogenic changes, including pollution and depletion, which were not intended to displace people.

The challenge for responding to climate-induced displacement is in determining where on the continuum between involuntary and voluntary migration the obligation to protect begins and ends. It is clear that those who fall at the "involuntary" extreme are due protection (although people displaced by "expropriation" would not fit into a climate change schema). But at what point as you move toward the centre of the continuum do protection entitlements/obligations cease? Bates's model, helpful as it is, does not offer a way of resolving this. Indeed, "deterioration refugees" seem in many instances also to be "environmental emigrants." Bates recognises this, noting that "connections between gradual environmental change and migration are rarely direct."[81] She also writes that whereas the other categories of "environmental refugees" have limited control over whether or not to migrate, environmental emigrants or deterioration refugees can "determine how they respond to environmental change."[82] It is precisely at the point on the continuum where "deterioration refugees" or "environmental emigrants" exist that any climate change protection regime must somehow draw the line.

Consider the situation of Suitupu Setema. He is a twenty-two-year-old from Vaitupu who sports three piercings in one of his ears; instead of earrings, he has inserted small splinters of wood. Suitupu arrived in Tuvalu's capital, Funafuti, a couple of years ago to pursue his studies and now works as a radio technician in the Tuvalu government's Media Department. He had observed coastal erosion and higher water levels on his home island, but it was while covering the local parliamentary sessions for his work that he learnt more about climate change. And although he cherishes his life in Tuvalu, he believes that the country will become uninhabitable. "I'd like to stay in Tuvalu because of the life and the freedom," Suitupu told me. "It looks like paradise. But I care about my children's future. They say that in the future, in one hundred years or fifty years, the water is rising."

So Suitupu is planning to migrate from Tuvalu to Australia or New Zealand. "There is no place to go [in Tuvalu]. There is no place to move. So it is better to make good plans for the future. I have to improve my education level, to get a degree or masters to easily get a job when I move to another country," he said. "The main reason I want to migrate is because of climate change. In a few years to come, we will have no land to stay in because the level of the water is increasing year by year."

Suitupu is not alone. He said that he has friends who are also planning to migrate. His friends discuss when might be the best time to move to another country. Others say that if people move, then there will be no one left. Older people say that you have to die for your country. But Suitupu said that he has friends who want to know how to migrate; they are planning to get a good job and a better education in order to improve their chances.

•

Suitupu is not in need of "protection." Yet his decision to migrate from Tuvalu is based, according to his own account, on his analysis of the impact of climate change on his homeland. A "push" factor, the prospect of his homeland becoming uninhabitable, is what motivates Suitupu to emigrate. Less vital is the "pull" factor of economic advancement – although this is clearly important, since Suitupu is the only son in his family and has obligations to support his parents in their older age. He has, or would like to have, choices. He would like to build his personal "capital" in order to emigrate from Tuvalu.

If he cannot accumulate this capital and migrate, and if his homeland does become uninhabitable, then Suitupu may be in need of protection in the foreseeable future. When this will occur and under what conditions is unknown. The IPCC has said, "It will be important to identify critical thresholds of change beyond which there may be collapse of ecological and social systems on atolls."[83] This is not merely a matter for curiosity. The collapse of these systems may not only be irreparable, but also potentially violent and traumatic.

This means that there is a case for programs whose focus is less on "protection" in an immediate sense and more on equipping those at risk of displacement with the social and economic capital to allow them to adapt and, if necessary, to migrate before they need protection. Such "pre-emptive" programs would be different from but complementary to a new protection regime. They would be designed to limit the need for urgent protection responses, to the extent that such needs can be predicted. Given what we know of the social and psychological impact of forced migration, pre-emptive adaptation and migration schemes may be more humane and potentially less expensive than finding protection solutions. Yet who would fund such schemes? What countries would be prepared, if necessary, to open their borders to allow people like Suitupu and the scores of others who may be in his situation to become members of their national communities? And what might Australia's obligations be?

CHAPTER 4

A neighbourly response

It is clear that climate change is of major concern in areas such as the economy, rural areas, health care, security, foreign relations and immigration. Climate change overarches all.

– PRIME MINISTER KEVIN RUDD,
2020 SUMMIT CONFERENCE, APRIL 2008

"Plane day" is an important event on Funafuti. It is when families, including those of the merchant seamen, that have been separated – for weeks or months or years – are reunited. There is a sense of excitement in the air. People mill around expectantly. Bead and shell sellers set up stalls in the shade under a great breadfruit tree adjacent to the airport. And there is the question that keeps everyone's mind focused: will the plane actually arrive today?

When planes are not using the airstrip, a black scar scratched into the narrow islet, it is a multipurpose community facility. There are no fences to quarantine the tarmac and it becomes a place for ball games, for gathering, for courting and for sleeping. Space is at a premium in Funafuti and overcrowded accommodation is common. So when there are simply too many people to fit into a house or when it is too hot to stay indoors, people head for the tarmac to sleep.

Semese Alefaio and I sit on couches in the Filamona's restaurant overlooking the airstrip. We are separated from the Tuvalu International Airport building by a narrow road and a *falekaupule* – a village hall. The tarmac is barely fifty metres from us. When planes touch down or take off, they thunder past the Filamona, silencing every conversation and halting every thought.

Semese speaks slowly, which is lucky, because I am scribbling notes and trying to keep up with him. He asked me not to tape our interview. The recording device makes him nervous, it seems. But Semese is not pacing himself just to help me. He speaks cautiously, judiciously. He doesn't answer questions immediately, but digests them first. He thinks about what he wants to say. He is not a political firebrand shooting off at the mouth.

Which is why it is confronting when Semese says that he is angry and frustrated by the plight of Tuvalu. It is unfair, he says, that Tuvalu has contributed so little to the problem of climate change and yet faces a disproportionate burden of its consequences. He feels "left out," like he is somehow an "alien" on a planet in which big countries pump out pollution that could ultimately mean the end of his homeland.

Semese is not alone in this sentiment. Many Tuvaluans realise that it is the wealthy and big developing countries that are responsible for climate change. Some know as well as many Australians do that Australia's response to climate change has been far from impressive.

During the years of the Howard government, Australia's response to climate change, detailed by Guy Pearse in his book *High & Dry* and by Clive Hamilton in *Scorcher* (both of which inform the following paragraphs), oscillated between denial and outright deception.[84] It is well known that the former prime minister and many of his senior colleagues were climate change sceptics. It is difficult not to conclude that this was either an ideologically driven conviction or political expedience. Even when Howard became convinced that anthropogenic climate change was an issue in mid-2007 he seemed to view the necessary responses as a potential threat to Australia's economic interests. The Australian economy, the government argued, was uniquely reliant on fossil fuels, and therefore would be hit particularly hard should it be required to cut its greenhouse gas emissions too dramatically. Further, to commit to meaningful cuts while developing countries such as China and India were not required to do so would be akin to handing Australia's economic rivals a competitive advantage. The Howard government also suggested that because Australia's emissions were small in global terms, a concerted effort to reduce them would reduce global emissions only marginally while disproportionately hurting Australia's economic interests.

Yet it was less the nation's economic interests and more the sectional interests of the big emitting industries that the Howard government protected. The self-styled "greenhouse mafia" – including representatives of the coal, mining, steel and aluminium smeltering sectors, and their lobbyists – was uniquely positioned to influence government policy. It used its access to power to prevent any movement on Australia's part towards lower emissions. And all the while, the government and the big emissions industries told the Australian public how successful they were in responding to global warming.

It is this power dynamic that contributed to the Howard government's refusal to sign the Kyoto Protocol, even after the former environment minister, Robert Hill, had extracted concessions from the international community. Under Kyoto, while the rest of the developed world committed to emissions reductions compared against 1990 levels, Australia was allowed emissions increases. Yet Howard refused to ratify the agreement and, in doing so, gave the largest emitter of greenhouse gases at the time, the United States, an ally in snubbing the international effort.

Australia's tardiness over Kyoto was mirrored in its dealings in the Pacific. Australia blocked and baulked and interfered at every chance. Its position on climate change dominated regional cooperation throughout the Pacific and undermined the region's ability to speak with a unified voice on the issue. "Australia's position was everywhere," according to the secretary to Tuvalu's Department of Foreign Affairs, Enele Sopoaga, "and if you look at it… it was almost like the extension of the US foreign policy. And it created a lot of very bad feelings for the Pacific. We didn't have even a regional stand on climate change. Climate change was bulldozed out of the [South Pacific] Forum agenda all these years of Howard. It is the most serious concern for the Pacific, and yet it was being pushed off the agenda... We were totally disorganised relative to the Caribbean region and the Indian Ocean because of Australia's position."

The Howard government's position on climate change helped to shape the nature of the public debate in Australia and was buoyed by its ideological bedfellows in the media. The few scientific voices that remained sceptical about climate change were given an undeserved prominence; the issue was presented as a debate between equally credible interlocutors, while the reality is that an overwhelming scientific consensus recognises the existence of anthropogenic climate change. All of this denial, deceit and obfuscation allowed the government to sidestep Australia's contribution to the problem of climate change, namely that Australia is the highest per capita emitter of greenhouse gases in the OECD and among the highest in the world.[85]

•

The Howard government's response to the threat of climate-induced displacement was no less inadequate. It partly reflected the government's contempt for the Pacific. In 2001, as we've seen, the media reported that a senior Tuvaluan official had requested that Australia and New Zealand allocate places for Tuvaluans to come to those countries in the event of climate-induced displacement. Reports indicated that while New Zealand had given in-principle support, Australia had rejected the proposal outright. On the face of it, there is no reason why Australia should have responded any differently; its hesitation may have been based on rational policy analysis. But the denial of Tuvalu's appeal, combined with the request of the Howard government a month later for Tuvalu to accommodate asylum seekers that Australia did not want to process within its own territory, suggests a contempt for Tuvalu that is consistent with the government's scepticism on climate change and its policy towards the Pacific.

The Howard government moved at every opportunity to prevent the question of climate-induced dislocation from being discussed. In 1997 it cajoled the South Pacific Forum into watering down a resolution dealing with "environmental refugees," threatening to reconsider Australia's aid allocations if its desires were not met.[86] And in 2001 Australia insisted that a reference to "environmental refugees" and to compensation for those affected by climate change be removed from a statement to be sent to the World Summit on Sustainable Development the following year.[87]

Closer to home, the Howard government appears to have been successful in promoting a sort of self-censorship among its bureaucrats on the question of climate-related migration. When a former CSIRO scientist, Dr Barrie Pittock, produced a manuscript for a book on climate change, his publisher, the Australian Greenhouse Office, recommended he remove a reference to the immigration consequences of climate change because it would be easier, without the reference, to have the book signed off by the relevant government departments.[88]

Furthermore, and consistent with the government's scepticism regarding the broader issue of climate change, there are indications that the government simply did not take the question of climate-induced displacement seriously. When asked about the Department of Immigration's planning for this displacement at a Senate Estimates hearing in May 2007, senior departmental officials told the committee that the department monitors the literature on climate change but had no contingency plan for the sort of serious displacement predicted in a number of studies.[89] Indeed, officials seemed to downplay the potential for large displacements affecting Australia. For example, the acting deputy secretary of the department, Peter Hughes, emphasised the possibility of internal as opposed to international migration resulting from climate change, saying, "It is not a necessary conclusion that international migration would be the direct consequence of climate change because, for example, in many circumstances an internal movement within a country – depending on the size and nature of the country – would be a solution, as opposed to international migration."

There is some truth to the assertion that the effects of climate change may be met in some or even many instances by internal migration – although as the Tuvaluan case demonstrates, not where internal migration is impossible. But Hughes's statement fudged the issue and suggested that if forced migration remains within the borders of a state, Australia has little about which to be concerned. His comments reflect the most optimistic of positions and underestimate the potential for these internal movements to blow out beyond borders. They also appear to indicate that the internally displaced are of no particular concern to Australia.

Departmental officials also emphasised that currently there is no international convention for dealing with climate-related displacement. As we've seen, this is true, but it is not an excuse for failing to grapple with the probability of displacement. It is incumbent on governments to seek to understand all manner of possibilities – both threats and opportunities – that may not yet have been articulated clearly in international forums.

It is possible that the lack of engagement expressed by senior public servants reflected something of the views of their political masters. In the Senate hearing Senator Ellison, Minister for Human Services at the time, sought to draw links between climate change and the drought cycle in Australia. "We have had our own droughts here," he said. "Then you have a good season and people are able to recover." In other words, we ought not to get too strung up about the potential for massive numbers of people being forced to leave their homelands due to climate-induced famine, drought or flooding because it is likely that they will be able to return to their homes at some time in the future. This ignores the fact that we are dealing not just with another weather cycle, but with an altogether unprecedented and potentially catastrophic set of circumstances. And it discounts the probability that in some instances, including in Tuvalu if predictions are accurate, return migration will simply be out of the question.

The Howard government's approach was also entirely consistent with its disdain for people who are forced to flee their homelands and arrive in Australia without first seeking permission. Throughout its term, the government implemented a set of policies and practices – including expanding mandatory, non-reviewable, indefinite detention, temporary protection visas for asylum seekers found to be refugees, and the Pacific Solution – that were designed to dissuade asylum seekers from coming to Australia in search of protection.[90] In the process, it did terrible harm to large numbers of asylum seekers, many of whom were later recognised as refugees. It also pursued a political strategy of demonising asylum seekers and refugees, casting them as different from "us" and somehow not fully human. For its own political ends, the Howard government promoted a mean-spiritedness within the Australian community towards asylum seekers and refugees. For a time, the strategy worked for the government.

Towards the end of his eleven years as prime minister, things began to change for Howard. The social researcher, Hugh Mackay, observed an awakening on the part of the electorate to issues of political morality that many had previously been content to avoid.[91] Australia's involvement in the war in Iraq was one such issue. Asylum seeker policy might have been another. After the emergence of a number of political scandals involving the Immigration Department and a change in sensibility within the electorate, the asylum seeker issue had lost its political virulence.[92]

At the same time, the electorate was being roused to the reality of climate change. It was a curious thing: while they had not seen clearly through the lies and deception, somehow people had come to be concerned about an issue that the government and powerful corporate interests had gone to extraordinary lengths to play down. According to political scientist Judith Brett, climate change was one of those issues that Howard's leadership style could not deal with effectively.[93] For Howard, climate change, like all political issues, was a contest between established ideological enemies. He could not understand that climate change was not an issue that divided the world like that. Instead, it was one that transgressed ideological camps and about which his "battlers" were becoming increasingly concerned.

•

Labor did not win the 2007 election because of climate change. But it did offer a more serious response to the issue than the Coalition and this helped to portray Labor as a party with a different approach. In opposition, Kevin Rudd had committed Labor to a 60 per cent reduction in greenhouse emissions by 2050 based on 2000 levels. The media was sceptical about how a Labor government would reach this target. Rudd maintained that he was waiting for the findings of a study, led by economist Ross Garnaut, that he and the state Labor governments had commissioned into the ways in which Australia could reduce its emissions.

The Rudd opposition had also committed to signing the Kyoto Protocol. Within hours of being sworn in as prime minister, Rudd moved to honour this pre-election commitment. Shortly afterwards, and amid much fanfare, Rudd and other senior ministers turned up in Bali for the latest international climate change conference. Some environmental groups and right wing commentators pilloried the new government for its stance at Bali. For some, Rudd did not go far enough in committing to emissions reductions; such people sided with the Europeans, who were advocating that developed nations sign immediately to binding emissions reduction targets of 25–40 per cent by 2020. Rudd argued, accurately, that the Bali conference was about setting out a framework agreement that would lead to binding emissions targets.[94] He also argued, as he had during the election campaign, that he would wait until the Garnaut Review was completed before committing to targets.[95] As for the criticism from the right, it was the same as usual.[96]

In the months leading up to the release of his draft report, Professor Garnaut indicated to the Australian community that the situation was worse than he had expected.[97] He suggested that deep cuts were going to be required and indicated his intention to design a broad-based emissions trading scheme. Sections of the community, most notably big emitting industries, lobbied to be excluded.

In the meantime, the climate change minister, Penny Wong, insisted – even in the face of Garnaut's alarming preliminary findings – that the government had promised a 60 per cent emissions reduction and that the government was not going to move.[98] It seemed that the Rudd government might be attempting to distance itself from the independent review it had hidden behind when asked, in the lead-up to the 2007 election, to provide details of how it would cut Australia's greenhouse emissions.

In May 2008, a stoush about the rising cost of fuel morphed into a debate about climate change. The Liberal–National opposition claimed that the government ought to decrease the cost of fuel by cutting the federal excise on fuel by five cents per litre.[99] Instead of standing firm and making the plausible case that high fossil fuel costs due to international factors are a reality of the future and that we may as well get used to it, the government capitulated, with Cabinet adopting the notion of "bowser parity," meaning that motorists would be compensated for any increase in petrol due to an emissions trading scheme.[100] The opposition claimed that petrol should be exempted from any scheme. A few weeks later a federal by-election in the Victorian coal-mining electorate of Gippsland resulted in a predictable defeat for the Rudd government. The pundits claimed that the electorate's sentiment reflected the government's insensitivity to the rising cost of fuel.

Brendan Nelson, the leader of the opposition at the time, continued to pursue a populist stance on climate change and emissions trading. "I suspect there's a high probability that we won't support what the government chooses to do," he said.[101] Nelson indicated that the poor would be hit hardest by emissions trading and that the government was rushing its implementation.

The much-anticipated Garnaut *Draft Report* was released on 4 July 2008. It asserted that by the end of the century, if nothing is done to mitigate climate change, the Murray–Darling Basin would be almost barren, the Great Barrier Reef destroyed and Kakadu damaged. It claimed that without action, Australia's annual national economic output would decline by 4.8 per cent over that period, and real wages would fall by 7.8 per cent. Garnaut argued for Australian action on climate change and for an emissions trading scheme to be phased in from 2010 – regardless of whether or not other members of the international community do the same. Under the Garnaut plan, money raised by a trading scheme would be used to compensate businesses and households, and be invested in research.[102]

The response to Garnaut's *Draft Report*, in light of the scientific evidence, led Australia's leading public intellectual, Robert Manne, to write that he felt as if he were living in a "parallel universe."[103] Neither the government, nor the opposition, nor powerful sections of the community seemed to grasp the seriousness of the threat posed by climate change and the need for a radical response merely to begin to avert the worse possibilities. The opposition insisted that petrol be excluded from any emissions scheme and then toyed with the idea of scrapping even the Howard government's policy, suggesting that Australia ought not to have such a scheme without major emitting nations being part of an international emissions reduction effort. "For Australia to go it alone, well in advance of the rest of the world, will do irreparable damage to our economic future and not do a darn thing to address climate change and an environmentally sustainable future for the planet," Nelson said. "It will be an act of environmental suicide, an act of economic suicide."[104] The position outlined by the leader of the opposition translates into a world in which all states do nothing to curb their emissions, preferring instead to wait on the actions of others and, in the meantime, cash in on any advantage to be made by continuing to develop emissions-intensive activities. Such an approach, while theoretically advantageous to a particular nation (although Garnaut challenges the economics of such an approach, arguing for the advantages of going early to a less emissions intensive economy), would have disastrous implications for humanity as a whole.

For its part, the government downplayed the significance of Garnaut's *Draft Report*, saying that it was but one of "a number of inputs" for its policy development on climate change.[105]

In the media, those who had run the line of the previous government and their big emitter allies, people such as Andrew Bolt[106] and Matthew Warren,[107] the *Australian*'s environmental reporter (a former public relations staffer for the NSW Minerals Council[108]), persisted. The ideological nature of the political right was evident in articles by the *Daily Telegraph*'s Piers Akerman, who called Garnaut's *Draft Report* "a fear-mongering document designed to bolster the old socialist agenda of wealth redistribution,"[109] and an editorial in the *Weekend Australian*, which welcomed the report but cautioned against "a return to left-wing social-engineering."[110]

Within two weeks of the report's release, the government introduced its own green paper on an emissions trading scheme.[111] It proposed a broad-based scheme, including fuel, but with concessions to major greenhouse gas emitting industries. And yielding again to populism, the government offset the cost of including fuel by promising to reduce fuel excise relative to any rise in cost to fuel associated with the scheme, at least for its first three years of operation. In essence, it was nullifying any reduction in fuel consumption, and therefore emissions, that might otherwise have come about through this element of the scheme. The opposition was not content with the fuel concession, arguing that the promise to review this provision in three years was cheap politics.

The Garnaut Review released its supplementary draft, *Targets and Trajectories*, and then its *Final Report* in September 2008.[112] It suggested that ideally Australia should participate in a global effort to reduced greenhouse gas concentrations to the equivalent of 450 parts per million (ppm) of carbon dioxide. This would mean that Australia would have to cut its greenhouse gas emissions by 25 per cent against 2000 levels by 2020 and by 90 per cent by 2050. Such a target would reduce the likelihood of dangerous climate change to 50 per cent. But the chances of an international consensus aimed at the equivalent of 450 ppm carbon dioxide are slight, according to Garnaut, so he recommended that Australia commit instead to an emissions target of 10 per cent (based on 2000 levels). If part of an international effort, this would result in a global greenhouse gas concentration of 550 ppm – a level that would see the Earth's climate tipping in dangerous and unpredictable ways. Should Australia "go it alone" because no international agreement is reached, Garnaut recommended a 5 per cent reduction in its greenhouse gas emissions.[113]

In light of the strength of the evidence Garnaut compiled about the dire consequences of climate change, these were extraordinary numbers for him to recommend. Even his most ambitious target – a 25 per cent reduction in Australia's greenhouse gas emissions by 2020 in order to minimise the risk of dangerous climate change – was at the lower end of the 25–40 per cent range agreed to by many developed nations at Bali earlier in the year,[114] making his pessimism about an international agreement seem like a self-fulfilling prophesy.

At the same time Garnaut's report reflected a sort of political realism, including a substantial dose of pessimism, as well as something of the parallel universes in which the climate science and the debate in Australia take place. The opposition, now led by former Howard government environment minister, Malcolm Turnbull, didn't challenge the government to commit to the ambitious emissions reduction figures that reflected the seriousness of the issue; instead, it maintained the position that an emissions trading scheme would be complex and costly and that the government was demonstrating undue haste in seeking to have one implemented by 2010.[115]

The government was attempting to balance apparently irreconcilable concerns. It conceded the need for significant emissions cuts, although the enviroment minister, Penny Wong, continued to insist that government policy was a 60 per cent reduction in Australia's emissions by 2050.[116] But the government also knew that such cuts will come at an economic and political cost. By September, the Garnaut process, which initially seemed to confound the government, was looking more like a political lifeline. Yet the government's lack of assertiveness was completely inconsistent with the increasingly compelling scientific consensus about the far-reaching action necessary to avert dangerous climate change. The cautious political approach on the part of the government was also out of step with the widespread belief in the community that the government has a mandate to act decisively on climate change and with polls that indicate that the majority of Australians would be prepared to pay higher costs to combat climate change.[117] The Rudd government had the opportunity in mid-2008 to lead Australia and the world in responding to climate change. By September 2008, while it had not yet finalised its position, indications suggest that on this important issue it was set to fail.

•

The Labor government's position on climate-induced displacement was indicated before the election in its discussion paper on climate change in the Pacific, prepared by the shadow overseas aid and Pacific Island affairs minister, Bob Sercombe, and Anthony Albanese, the shadow environment minister. The paper, *Our Drowning Neighbours*, contained some exaggerations, including the assertion that Tuvalu "is expected to become uninhabitable because of rising sea levels over the coming decade, with its entire population having to relocate to other countries."[118] But the paper did show an intention to engage more meaningfully with the Pacific on climate change, proposing a Pacific Climate Change Strategy. Such a strategy, as well as containing provision for monitoring and assistance for mitigation and adaptation, should specifically address the question of displacement, according to the paper. It should include "[a]ssistance with intra-country evacuations when citizens have to be moved from low-lying areas to higher ground" and establish "an international coalition to accept climate change refugees when a country becomes uninhabitable because of rising sea levels."[119] As part of this international response the paper suggested that "Australia should be working at the UN to ensure appropriate recognition of climate change refugees in existing conventions, or through the establishment of a new convention on climate change refugees."[120]

Within six weeks of taking office, the Rudd government had dispatched two of its members, the parliamentary secretary for Pacific Island affairs, Duncan Kerr, and the parliamentary secretary for international development assistance, Bob McMullan, to the Pacific. Part of their job was to investigate the impact of climate change on the region.[121] Both the seniority of the government members and their presence in the Pacific were powerful signals of the Australian government's willingness to take seriously the Pacific and the challenges it faced. Pacific Island nations indicated a keenness to work with a new Australian government less inclined to play the "tough cop" of the region.[122] Tuvalu has noted Australia's changing attitude. "I am optimistic, but it is too early to tell," Enele Sopoanga told me, adding that "this new Australian government is very practical and I think they will be serious about their engagement with the Pacific and I think they will be very, very practical and I am very optimistic that maybe we can do a lot on climate change."

The changed tone has also been noticeable in parliament. In a Senate Estimates hearing in February 2008, just a few months after taking office, the immigration minister, Chris Evans, was asked by Greens senator Kerry Nettle whether "climate refugees" were a priority for the government. The minister acknowledged that the issue had not been his main concern in the short period of his tenure; he had been left with a good deal to clean up, including ending the Pacific Solution, resolving a couple of hundred cases of long-term detainees and finding a way of normalising the status of thousands of refugees on temporary protection visas. But, in a manner completely different from that of the previous government, he went on to say: "I am conscious it is an issue that may confront the government. The government is focused particularly on the South Pacific strategy and supporting those islands and communities in our near neighbourhood. Of course, some of them are right in the firing line for any climate change impact, so it is a policy issue confronting the government. I have not taken any specific measures in relation to immigration, but I am well aware that it is an issue for our portfolio and there has been discussion among government, particularly about the South Pacific issues and how that might impact et cetera."[123] The minister also told parliament that "climate refugees" were on the agenda for a ministerial council meeting which involved state ministers for multicultural affairs and a New Zealand government representative.[124]

The Rudd government has allocated $150 million from its aid budget over three years for climate-related adaptation.[125] This is consistent with Australia's preference for adaptation and "well supported internal relocation" over resettlement as a response to climate-related displacement.[126] It is also consistent with the established norm in refugee protection, namely that resettlement in countries like Australia is the least preferred "durable solution" to refugee movements. Yet the preferred solutions – repatriation to refugees' countries of origin or local integration into countries of first asylum – may be entirely untenable in some instances of climate-induced displacement. For example, repatriation may be impossible because there may be no habitable country to which to return. And local integration may not only raise important issues of global equity – the vast majority of refugees flee poor countries and seek protection within similarly poor neighbouring countries, a trend that is likely to continue in instances of climate-induced international migration – but also of sustainability, especially if, as predicted, millions of migrants move from a place that has become uninhabitable to another that is only marginally habitable.

Notwithstanding the Rudd government's more positive rhetoric and its investment in adaptation, its response to climate-induced displacement remains slow. Chris Evans recently spoke of Australia's continuing engagement with other countries on this issue, and of Australia's "flexibility to assist with international resettlement of people displaced by climate change if, at some point in the future, this becomes appropriate." It seems that the government is prepared to pursue an *ad hoc* approach to this question if and when it becomes pressing. In early September 2008, contrary to the Labor Party's pre-election discussion paper, the immigration minister's media adviser told me that the minister is not engaged in any international discussions about the appropriate recognition of those displaced by climate change, either in existing international conventions or in a new instrument designed to respond to climate-related displacement.[127]

Further, the substance of the bureaucratic response is not altogether different from that under the previous government. There has been a change in tone on the part of Immigration Department officials, reflecting, it would seem, the views of their new political bosses; for example, the more abstract "monitoring the literature and studies on climate change for the people movement dimensions" had become a research project involving a survey of the literature on climate change, displacement and the Pacific.[128] But at the Senate estimates hearings in May 2008 the department continued to echo its earlier response, simply noting that there is no specific regime for dealing with people displaced by climate change and that it holds discussions with immigration officials from other countries.

•

Migration is a complex policy area. It involves not just the act of physically leaving a homeland but a host of social, political, economic and environmental causal and precipitating factors. The most significant way to respond to refugee movements is to understand and address these root causes and to eliminate the reasons for persecution and human rights violations. This an extremely difficult task.

Responding to climate-induced displacement also involves addressing root causes. It must involve mitigation and adaptation. It will demand changes in the way we structure our economies and live our lives. It will also require well-considered aid and development assistance to help local and national communities to build the capacity to respond to the effects of climate change. These are ways of addressing the root causes, if not of climate change, then of the dislocation that may stem from it. The Rudd government's interest in a Pacific labour migration scheme, similar to a New Zealand scheme, may be one way of dealing with underdevelopment; remittances, estimated to be worth more than international aid and development assistance worldwide, may be one way of promoting development.[129]

But there are also times when root causes cannot be addressed. These cases call for an international framework and for domestic policies and practices – intellectually rigorous and yet compassionate – to respond to the displaced. At a domestic level, there is a need to incorporate climate-related displacement into Australia's humanitarian immigration program. Simply claiming that Australia's migration program has the flexibility to respond to climate-related displacement is insufficient. There is a risk in viewing such displacement merely as a general migration issue rather than a particular sort of migration, namely protection-related migration. This distinction matters because people who are displaced by climate change may well have particular needs that are the result of their experiences of upheaval that other migrants do not – in the same way that that the refugee's experience calls for specific resettlement services. Other countries, while not specifically addressing climate-induced displacement, have made provision for people dislocated by environmental factors. Sweden, for example, has a specific migration category of "person in need of protection" due to natural disasters. And in 2003, the US extended Temporary Protection Status to 80,000 Hondurans who had fled to that country after Hurricane Mitch destroyed parts of central America in 1998.[130]

If the projections of climate scientists turn out to be accurate, there is the real possibility that climate-induced dislocation could become a major international issue in coming decades. This could have far-reaching implications for the world and for Australia. It is in our national interest to respond to climate change and to the displacement that will result from it. This notion of the national interest has long been central to Australia's immigration and refugee program. In the 1940s and 50s, mass migration to Australia was encouraged in order to meet Australia's security and nation building interests.[131] Even when Australia first articulated a distinct response to refugees – as opposed to viewing refugees merely as migrants – in 1977, the national interest remained dominant.[132] John Howard's slogan in 2001 – that "we will decide who comes and the circumstances in which they come" – was a reminder of the importance of state control over migration to Australia in the national interest. Clearly, the national interest will be a central consideration in responding to climate-induced displacement.

But the national interest ought not to be understood narrowly. It ought not to be used as a reason for closing our borders to those seeking protection from climate-related factors any more than it should be used to close our borders to refugees. It will not do for Australia to face a world in which tens or hundreds of millions of people are on the move because of climate-related factors and claim that it is in our national interest merely to guard our sovereign territory. Rather, it is in our national interest to be engaged with the world to find solutions to climate change and the displacement it is likely to cause.

It was in Australia's national interest to be engaged, albeit not altogether positively, in the drafting of the 1951 Refugee Convention.[133] It was also in the nation's interest to develop effective responses to Indo-Chinese refugees from the mid-1970s and later, to participate in the United Nations-sponsored Comprehensive Plan of Action which, although imperfect, was implemented to resolve the ongoing fallout from the Indo-Chinese refugee crisis.[134] It was in Australia's national interests to respond to these issues not only for security or economic reasons, but also in order to have an influence on the outcomes of processes that would ultimately impact on Australia. It is also beneficial for Australia to be seen to be an engaged international citizen. In the case of climate-induced displacement, being a good international citizen might be understood as participating in seeking solutions that are not only effective in protecting those in need of protection, but which make a contribution on the basis of principles of justice. Our contributions ought to take account of our comparative wealth – Australia is one of the wealthiest countries in the world, having achieved this wealth through activities that emit greenhouse gases – and our contributions to the problem of global warming given that we are the highest per capita emitters of greenhouse gases in the OECD and the fifth highest in the world. Like Australia's response to previous refugee crises, responding to climate-induced displacement will involve Australia's immigration program as well as our active participation in international efforts aimed at addressing the problem. And we have a good deal to offer the international community: Australia is a country for which immigration has been central to social and economic development, which has a successful refugee resettlement program and is an active contributor to international forums concerning refugee protection.

CHAPTER 5

Brave new world

This century will be more and more a century of people on the move, for all sorts of reasons – economic reasons, also climate change, war and conflict – and in a sense all societies will tend to be multi-ethnic, multi-religious, multicultural, and people will need to live with each other and respect each other... [W]e live at a time when there is a risk of moving to a post-Enlightenment society where the values of reason and tolerance of the Enlightenment... can be extinguished.

– United Nations High Commissioner for Refugees, Antonio Guterres, interview with Frances Williams, FT.com, 2 January 2008

The international scientific community has told the world that in order to prevent catastrophic global warming we need to curb greenhouse gas emissions. By 2007, the concentration of carbon dioxide in the atmosphere had risen to 383 parts per million compared against pre-industrial levels of 280 ppm.[135] This is just carbon dioxide. In 2005, long-term greenhouse gases in the atmosphere – including carbon dioxide and other gases – were equivalent to 445 parts per million of carbon dioxide.[136] According to the Garnaut review, a stabilisation target of 450 ppm CO_2 equivalent would give about a 50 per cent chance of limiting global mean temperature rise to two degrees above pre-industrial level and thereby averting the risks of dangerous climate change.[137] Some scientists believe that this figure is too high.[138] Higher levels of emissions would mean an increased likelihood of catastrophic climate change.

The 2007 Intergovernmental Panel on Climate Change report presented overwhelming evidence that climate change due to human activities is real and that its effects may well be profound. Compelling as it was, the panel's report was dated even before its release. It did not make reference to the more rapid than expected melting of the Arctic ice cap, revealed in September 2007, which indicated that the extent of the thaw was one hundred years ahead of what had been expected.[139] According to one scientist, there may be no Arctic ice cap in summer within a decade. This could have further feedback implications, increasing the "albedo flip" where the Earth absorbs even more heat because the white ice that reflects solar radiation is replaced by darker seas which absorb it.[140] The melting of the Arctic ice cap could also increase the rate at which the Greenland ice sheet is melting. Nor did the 2007 IPCC report include satellite data, released before the report, that showed that sea levels had increased by an average of 3.3 mm per year between 1993 and 2006 – above the 2001 IPCC's best estimate of less than 2 mm per year over the previous century.[141]

The implications of the increasing speed of climate change for a country like Tuvalu is obvious. The country is positioned precariously close to the surface of the sea. It is already experiencing sea level rises of potentially 5.7 mm per year. Locals say that sea level rise is affecting their traditionally grown food and water supplies. It is also contributing to unusual levels of coastal erosion, as well as to sea inundation and salt intrusion. Should the effects of global warming continue, it is likely that Tuvalu will become uninhabitable in the not-too-distant future. Exactly how this would occur is unknown. Will it be because of increased scarcity as local food and water supplies diminish and it becomes uneconomical to import all the nation's food? Will it be because one extreme weather event after another – events like Hurricane Bebe, which struck in 1972 – batters Tuvalu so that the risk of remaining becomes too great to bear? Will it be a combination of these factors? And what would the social implications of this deterioration of life be? The answers to these questions are all unknown. But the fate of 11,000 Tuvaluans rests on the actions of the international community. There may still be time to lower greenhouse emissions to a safe and sustainable level. There are certainly opportunities to provide Tuvalu with resources to build resilience against some of the effects of climate change.

Tuvalu is only a very small example of something potentially much larger. Resettling 11,000 displaced persons would be relatively easy. Each year since 2002–03 Australia has admitted more than 100,000 people as migrants.[142] A further 13,000 or so people who are in need of protection are resettled here, partly because Australia accepts that it has a role in sharing the burden for international protection, the vast weight of which falls on the poor, neighbouring countries who shelter most of the world's refugees.[143] It would be possible to absorb the entire population of Tuvalu in a single year's migration quota, if the need presented itself. But predictions indicate that displacement due to climate change will not stop at Tuvalu. Rather, it is predicted that tens or even hundreds of millions of people will be displaced by climate change by 2050. There are claims that some, including the people of the Carteret Islands who have already started to resettle in neighbouring Bougainville, are being forced to leave their homelands due to rising seas caused by climate change.[144] On its own, sea level rise could dislocate many millions. According to one recent study, a sea level rise of one metre would expose 117 million people to inundation in Asia alone.[145] This would include more than seven million Indians, nearly fifteen million Bangladeshis, more than seventeen million Vietnamese and some seventy-two million Chinese. Over thirteen million Europeans and over eleven million Africans, including eight million Egyptians and more than three million Nigerians, would be at risk of inundation.

These figures relate only to a rise in the sea level. Drought and flood will displace many across the globe. It is projected, for example, that the arid regions north of the Sahara will become hotter and drier due to climate change; the rain that would have been expected to fall on this region, according to scientific knowledge, will fall in future in the higher latitudes. This will lead to increasing water and food stress on those who live in the Sahel, potentially contributing to outward migration.[146] Then there is melting of permafrost in Arctic regions and severe weather events such as cyclones which will also displace large numbers of people. Much of this movement will be within the borders of states, but significant numbers of people, particularly in countries where there are few opportunities for internal relocation, will cross international borders. This could add to conflict within and between states. The most vulnerable people, because of their socio-economic status, will be women.[147]

Climate-related displacement is now on the agenda of the United Nations High Commissioner for Refugees. In June 2008, the commissioner, Antonio Guterres, said, "Climate change is today one of the main drivers of forced displacement, both directly through impact on the environment – not allowing people to live any more in the areas where they were traditionally living – and as a trigger of extreme poverty and conflict."[148]

•

The very nature of climate change, its scope and complexity, our limited understanding of the systems it affects, and the uncertainty of the human response, mean that its effects remain unpredictable. Because of this, the literature on the future effects of climate change, while grounded in science and in observed phenomena, remains speculative. The threats posed by climate change are largely future threats. The IPCC has developed a number of future scenarios based on four "storylines." These storylines present different future demographic, economic and technological possibilities in order to project levels of emissions to 2100.[149] In a similar vein, the Garnaut Review has developed its own case studies. But unlike the IPCC's scenarios, Garnaut develops three emissions cases – the "no-mitigation case," the "550 mitigation case," and the "450 mitigation case" – as a way of describing how a future climate might look depending on how successful we are at reducing emissions.[150]

Both the IPCC and the Garnaut case studies help us to understand the possible impacts of climate change, depending on the amount of greenhouse gas we continue to pump into the atmosphere. In a similar way, scenarios assist in understanding the human implications of climate change. This, of course, is far from an exact science. While it is possible to know with reasonable certainty that a sea level rise of one metre, without other adaptation, will affect many millions of people who live at an altitude at or less than a metre above sea level, it is not certain how these those affected – both as individuals and as communities – will behave in such circumstances. Furthermore, a sea level rise of one metre would take some years and would be accompanied by a range of other unpredictable environmental and social factors.

In a recent report, *Climate Change as a Security Risk,* the German Advisory Council on Global Change constructed a number of regional and state-based fictional "narrative scenarios" through which to explore the security risks to societies, nation-states, regions and the international system associated with climate change. These scenarios blended forecasts about changes in the Earth's climate with other socio-economic and political factors based on empirical research in order to "tell the stories" of possible future security developments.[151] The scenarios were developed as a means of identifying the security risks associated with climate change and the sorts of decisions that might help to avoid these risks. The narrative scenarios begin around 2020 and end around the middle of the century, with some extending to the end of the century. They presume, as the science has indicated, that the effects of climate change up to the middle of the century are largely set in place.

One of the scenarios the German Advisory Council presents is of Bangladesh.[152] Bangladesh is a country of more than 150 million people. Half of this number lives in poverty. The average annual income of a Bangladeshi is less than what an Australian will earn in a week.[153] Bangladesh is the most densely populated and one of the poorest countries in the world. It is also, due to its geography, one of the most prone to natural disasters. Bangladesh is home to the largest river delta in the world where three great rivers – the Ganges, Bramaputra-Jamuna and Meghna – meet before spilling into the sea. Its elevation above sea level is generally less than 10 metres, and in the delta region, far less than this. Furthermore, Bangladesh, located at the top of the Bay of Bengal and to the south of the Himalayas, sits right within a tropical cyclone region. Bangladeshis regularly experience floods, droughts, earthquakes and tropical cyclones. Each year, thousands are killed, dislocated and have their livelihoods stripped from them due to natural disasters. Each year, the costs of natural disasters adds up to several thousand million dollars, money that might otherwise be spent on poverty alleviation and development.

Now add climate change into the mix. According to the German Advisory Council a "plausible development" in Bangladesh by 2020 – that is, in just 12 years – is a 12 centimetre rise in the sea level compared against the 2000 level. This would mean the loss of residential land, fertile farmland and other coastal ecosystems that provide livelihoods for millions. Millions would be forced to leave their homes. The rise in sea level would lead to "catastrophic damage" caused by storm floods as waves batter and inundate the coast, salifying land and fresh water supplies. Food security and access to clean drinking water would be diminished, leading to under- or malnutrition and flourishing water-borne diseases.

Nor is this the worst of it. The German study provides a further "disaster" scenario in which it posits the consequences of a failure of the international community to develop and implement effective climate change mitigation policies by 2020. According to this scenario, Bangladesh would be hit by stronger and stronger cyclones, and by massive storm seas. Hundreds of thousands would be killed. Millions of people living on the coast would be internally displaced. The economy would stagnate. The legitimacy of the government would be undermined by its inability to deliver disaster relief and other services, leading to political and social instability and the rise of extremist movements. Tens of millions of people would be forced to flee into neighbouring India, thereby increasing tensions within that country and internationalising the potential for conflict.

The narrative scenarios for other parts of Asia, as well as Africa and South America, are of a similar flavour. The impact of climate change, combined with socio-economic and political factors will cause massive numbers of people to flee their homes. It is generally accepted that most of this movement will be within state borders or, when it crosses state borders, will be confined to the regions from which the displaced come. But even if a small fraction of one or two hundred million displaced persons seek to enter developed countries like Australia, this would equate to a large number of people. The unknown question, and one that is not addressed by the German Advisory Council, is: how will we, and other countries like ours, respond to this new displacement?

•

In the March 2004 Quarterly Essay, *Sending Them Home: Refugees and the New Politics of Indifference*, Robert Manne and I followed the lead of Isaac Deutscher and wrote of a fictional "Iranistaqi" refugee in an effort vividly to describe why refugees on temporary protection visas should not be repatriated from Australia.[154] I am repeating this exercise here both to begin to imagine the human reality of climate change–related displacement and as a way of pre-empting and thereby preventing one possible response on Australia's part to such displacement. This approach is consistent with the scenario-based approach to climate change more generally. It is, of course, an imaginative exercise.

It is 2020. Over the past decade, the fictional state of Sahesh, a developing nation of fifty million people, has been hit by extreme drought and devastating tropical cyclones. The seemingly constant environmental disasters have overwhelmed the government's capacity both to provide assistance to the stricken and to rebuild the country's vital infrastructure. More than 60 per cent of the population lives in poverty; they have inadequate shelter and no access to clean drinking water and are malnourished.

Mustafa is a thirty-five-year-old father of four from a middle class family of farmers in the north of the country. As a child, it was expected that he would continue in the tradition of generations, eventually taking over his father's land. But after years of trying to adapt to the unpredictable cycle of drought and flood, the result of global climate change, Mustafa and his family decided that he must join the wave of rural Saheshis searching for employment in the city. Mustafa left his wife and children with his parents, and found a room that he shared with eight other men in an urban slum on the far outskirts of Sahesh's capital.

Notwithstanding his high school education, Mustafa could not find steady employment. Instead, and like millions of others, he spent hours in queues hoping that it might be his turn to be picked for a day's work on a construction site. When he was fortunate enough to find work, he sent whatever money he could to his family. The rest he kept for his own living expenses and to pay off the local insurgency, because in the midst of the filth and poverty, a militant politico-religious movement had grown up, mixing an anti-corruption critique of the material poverty of the country with anti-western sentiment and a vision of religious salvation. It extorted money from the day labourers to fund guerrilla attacks on the government as well as a sort of privatised welfare net for those who could not meet their basic needs. While Mustafa could identify with the causes of the insurgency – was he not part of their "disenfranchised masses"? – he was unconvinced of their ideological critique and temperamentally disinclined to support their tactics. Yet, because of his financial situation, he had no choice but to live in a slum that was controlled by the movement and which was frequently attacked by the government as part of its counter-insurgency strategy.

In 2018, Sahesh was hit by a particularly destructive cyclone. Tens of thousands of people were killed, including Mustafa's youngest daughter and his father. Thousands of others were injured. Whole villages disappeared. The capital was devastated. As well as being battered by the winds and rain, the city was inundated by a wave surge. Buildings collapsed and roads washed away. The land around the port was swallowed up. The airport, on the sea side of the city, was destroyed. The damaged infrastructure delayed relief efforts, causing food shortages and promoting the spread of waterborne illnesses that led to the deaths of thousands more Saheshis.

Mustafa's wife, his surviving children and his mother fled their home to find shelter in a long-established camp for internally displaced people. The camp was under the nominal administration of the United Nations, but was controlled by one of the factions of the insurgency. It was was rife with crime and corruption. Women without menfolk, like Mustafa's wife, his surviving daughter and his mother, were particularly vulnerable to the predations of gangs of young men.

Mustafa decided that he needed to find a safer place in which to live and to raise his family. The absence of security and of a life with dignity meant that it was impossible to remain. He organised for his family's remaining livestock and his father's motorcycle to be sold. Then, through a housemate with a link to the insurgency, he found himself a people smuggler. (The insurgency maintained close links with organised crime because the crime rings imported weapons and luxuries like bottled water and tinned meat as well as exporting other commodities, including human beings.)

Unable to afford the fares to North America – which by then had erected a wall across its southern border that was patrolled by government-funded vigilante groups – or Europe, Mustafa set out for Australia. By foot and truck, by bus and boat, he travelled to Indonesia, a country with its own massive displacement problems, where he waited for a place on a boat with almost a thousand people seeking to make the journey to Australia.

By the time Mustafa arrived in Australia, hostility against migrants in general, and particularly against those seeking protection, had grown. Fuelling the shift in sentiment was the economic vulnerability of sections of the Australian community, caused largely by a downturn in the global economy due to soaring oil prices and the impact of climate change, felt particularly in agricultural communities, as well as a sense that the new Asia-Pacific Community had left many Australians disenfranchised within their own nation. Non-citizens seeking protection from persecution and climate change became symbolic, for many Australians, of the dangers the nation faced in an uncertain and changing world. Protection seekers were a manifestation of uncontrollable climate change on a global scale and a reminder of Australians' sense of powerlessness. They were also a threat to the Australian environment, which was struggling to provide for those who were already here. They were economic competitors and they were security threats. And, importantly, they were a threat to the way some people had come to see the nation – as a sort of homogenous, meaning-giving thing – and their superior place within it.

Rather than challenging this sentiment, Australia's prime minister used it to political advantage. On climate change, the prime minister said that Australia had already done its share. Indeed, according to the government, Australia had done more than its share: Australia was no longer the largest per capita emitter of greenhouse gasses in the OECD and had contributed to international mitigation and adaptation efforts – participating in an emissions trading scheme, funding adaptation projects in vulnerable countries. Now, so the argument went, Australians were hurting unfairly. It was time, as the government's spin put it, to "Protect Our Own." Many Australians agreed with the government that it was time to bunker down, to put Australia and Australians first. And Australia's response to people seeking protection became a potent symbol of doing so.

Although the Rudd government had abandoned it more than a decade earlier, the new Australian government used the Pacific Solution as the blue-print for its response to the next wave of "boat people." It sent out the navy to intercept and turn back the boats. It fired warning shots over the bow of Mustafa's boat and ordered it to leave Australian territorial waters. Later, the boat was deemed too unseaworthy to return to Indonesia. Mustafa was brought to Christmas Island, where he was kept in the detention centre, a state-of-the-art facility, completed in 2008, with a capacity to hold 400 people and contingency for another 400. In 2019, when Mustafa arrived, there were 2200 people, some having been there for more than three years.

Mustafa's claim for protection was assessed by the Immigration Department against the 1951 Refugee Convention and the more recent 2015 Protocol Relating to the Status of Climate Change Displacees. Initially his claim was rejected, so he appealed. But Christmas Island, under laws passed in 2001, was "excised," meaning that protection seekers who were brought there had fewer rights than anyone on the Australian mainland. Although the Rudd government amended the excision laws in 2008 to allow asylum seekers access to legal advice and to an independent merits review of negative decisions, the new government repealed these amendments so that the same people who had made the first decision in Mustafa's case were also the adjudicators of the review. Notwithstanding the lack of an independent review process, on the second attempt, Mustafa was found to be a Climate Change Displacee in need of protection. By the time the review decision had been made, Mustafa had been in detention for nine months.

But then he waited. The prime minister had told Australians that no protection seeker would make it to Australia henceforth. So Mustafa waited in detention for another nine months. His frustration and anger grew. Inside the detention centre, a culture of self-harm developed. Like all the detainees, Mustafa became depressed and hopeless. His desperate concern for his wife and children back in Sahesh was all-consuming, but he was helpless to act. He began to slash his body, an act that he knew was irrational but was also, somehow, a relief. He could not return to Sahesh; nor was he allowed to enter Australia. It was unlikely that he would be accepted for resettlement in Europe or North America; with a hundred million displaced by climate change added to the thirty million refugees and internally displaced people, they had their own immigration issues. Indeed, the situation in Europe was worse than Australia; while Australia had invented the Pacific Solution, the European Union had perfected it. Now, as millions of people from across Africa fled north in search of protection from drought and poverty and war, they faced a joint European Border Patrol Taskforce charged with preventing them from crossing the Mediterranean and entering Europe.

The best Mustafa could hope for was resettlement in one of the Reservations that had been created as part of the negotiations for the 2015 Climate Change Displacees Convention. In a perverse form of international "burden sharing," wealthier countries such as Australia could pay to have refugees and displacees housed in reservations located in the poorest parts of the world – in Papua New Guinea, Libya and Romania.[155] In return for sheltering the displaced, the countries in which the Reservations were located were paid handsomely. The Reservations were generally protected from the brunt of climate change – they were not subject to sea level rise, and the international community footed the bill for basic necessities such as fresh water and food to be airlifted in. In this sense, the Reservations did not amount to a violation of international protection obligations: sending refugees and displacees there did not amount to a breach of international *non-refoulement* principles. According to the Australian immigration minister, refugees and displacees who complained about the Reservations were not in need of protection but, rather, were seeking "migration outcomes." But the Reservations were scenes of hopelessness. Initially they had been conceived of as places of temporary refuge, places in which the displaced could be accommodated briefly while their homelands recovered from war or disaster or where they could be processed for resettlement in other countries. Later, as it became clear that the opportunities for resettlement in countries like Australia were entirely inadequate for dealing with the massive numbers of permanently displaced, the Reservations became overcrowded, squalid and unsafe. They became "warehouses" for the world's rejected. As one academic refugee specialist described it in the 1990s, reflecting on the wealthy world's efforts to keep the poor, coloured folk out, this was "global apartheid."[156]

•

This scenario may seem alarmist, but it is not altogether far-fetched. Indeed, it is based on Australia's response to asylum seekers over the life of the Howard government, attempts by European countries to follow Australia's lead, and real debates conducted by respected scholars in the academic literature about how the international community ought to respond to a refugee protection regime that is already insufficient.[157]

Australia's response to the threat of climate change–induced displacement need not turn out this way. But there is a risk that it will. At a domestic level, the key risk factor is politics: how will Australia's political leaders choose to confront the question? Will they use it as an opportunity to promote a populist and exclusive Australian nationalism or will they develop policy and political responses that are hard-headed but also humane and generous?

Internationally, there is a need for mitigation and adaptation measures – and a means of funding such measures – that will help to prevent displacement. There is also a need to establish a framework that sets out the rights of those displaced by climate change and the obligations – including financial obligations – that states and the international community have for such people. This will be a protracted process, and one in which Australia should be an active participant. Australia is situated in a region that is home to 60 per cent of the world's population – a region that will be affected most dramatically by sea level rise, if not other climate-related factors.

Clearly this is not just a matter for fictitious Mustafa. It is also a probability for those who live in the run-down fibro shacks painted in different colours, some with warped masonite window shades, some with louvred windows, in Funafuti. Even if they don't think about it, it is a possibility for the women sweeping the litter into piles in their coral-dust yards and for others who sit playing cards in the doorways of their homes. It will be the reality for the kids playing among the piles of coconut husks and the free-ranging chickens, or riding bicycles and pushing tyres along the road, and possibly for the men who sleep in open-sided sheds or in hammocks strung between trees. It is a matter that could well affect not only these unknown Tuvaluans, but also Solomina, Toma, Tasi, Semese, Mitala, Melete, Suitupu, Tui, Tuu and Luke.

It is also a matter about which Tufitu already has some insight. She remembers the fear of the dark hours of 17 April 2006 and the brief dislocation of her family. Yet she wants to stay in Tuvalu. "I don't want to leave," she said. "This is my home." But Tufitu also realises that, in the end, if Tuvalu becomes uninhabitable, she may have no choice but to leave. And she knows that if that is the case, it is a problem that is not of her making. There is a profound sense of injustice in the fact that countries that are wealthy or that are getting wealthy will be the cause of her displacement. As she sat near her family's homes she reflected on this. "I feel sad about it," Tufitu said. "Please stop [emitting greenhouse gases] – that is our plea. From me, I say, just stop it.

Don't create more. That may affect climate change and cause our country to suffer. I mean people: Just feel sorry about us."

From the "ocean side" of the islet, it is not difficult to feel Tufitu's vulnerability. I had been there a few nights before I met Tufitu. The sun was sinking over the lagoon behind me as I walked between the pig pens to the east of the airstrip. The sky over the ocean was tinged with pink. A full moon already hung in the sky. It was beautiful. A small swell was breaking out where the reef began, about twenty metres from the shore. The tide was full. When I had been to the ocean side before, the tide had been out. I had seen men fishing with their nets along the reef, and hermit crabs that snuck into their shells as I approached them. But this time, I was conscious of a sense of insecurity from being so close to the ocean and so far from any more substantial landmass. Even with the reef protecting the atoll from the full brunt of the ocean, the waves were coming to within twenty centimetres of the top of the bank. As each wave receded it rattled back along the coral stones, until the next wave climbed the bank.

The banks are the highest points on the island. It is easy to see how they could be breached. On this porous atoll, it is also easy to see how rising sea levels could encroach on the land from below.

Australia ought to be taking the lead in international efforts that develop a framework for responding to the possibility that Tuvaluans and many millions of others may be displaced due to climate change. To do so is not to concede that mitigation and adaptation efforts are beyond us, although the longer we delay, the more real this possibility becomes. Rather, planning for a future of mass displacement due to climate change gives us the opportunity – before millions of people are on the move throughout the world because of climate change; before we, and other nations, become tempted to erect walls to keep them at bay; before we start to say as though as a reflex that "we will decide who comes and the circumstances in which they come" – to develop frameworks and institutions that might not only be politically realistic, but also based on principles that promote human rights and dignity.

Acknowledgments

I am thankful to Peter Browne and Klaus Neumann for coming up with the idea for this book and for giving me the opportunity to write it. Their comments on various drafts and Peter's editorial skills made the book a better one. Of course, any shortcomings are mine.

In Tuvalu, I was assisted by Annie Homasi, Taukiei Kitara, Semese Alefaio and Meleka Tausi from TANGO. Luke Paniue was a great help, as was Tatuaua Pese.

Back in Australia, Richard Hanley and Dorothy Leslie provided space for me to write, as well as lots of conversation, coffee and the occasional salsa lesson.

My partner, Nadine, took care of Tenzin and Annie while I was away and freed me to do the work required to research the book. Again I thank her for her love and support.

Endnotes

1. Tuvalu News, "Brisbane Journalist in Tuvalu for Extreme High Tides," 13 February 2006, http://www.tuvaluislands.com; Ricardo Morris, 'Meteorologists Warn of King Tides to Sweep in Today,' *Pacific Magazine*, 28 February 2006.
2. IPCC, *Climate Change 2007: Synthesis Report*, http://www.ipcc.ch, p 30.
3. Ibid.
4. A Report of Working Group 1 of the Intergovernmental Panel on Climate Change, *Summary for Policymakers*, Third Assessment Report, IPCC, pp 2–4.
5. Ibid, pp 2–4.
6. David Spratt & Philip Sutton, *Climate Code Red: The Case for Emergency Action*, Scribe, Melbourne, 2008, p 33.
7. IPCC, *Climate Change 2007: Synthesis Report*, op cit, p 30.
8. Ibid, p 30.
9. For example, the concentration of CO_2 in the atmosphere has increased from 280 parts per million (ppm) in pre-industrial times to 379 ppm by 2005. This 2005 figure "exceeds by far the natural range over the last 650,000 years (180–300ppm) as determined from ice cores": IPCC,

"Summary for Policymakers," in *Climate Change 2007: The Physical Science Basis: Contribution of Working Group 1 to the Fourth Assessment Report of the Intergovernmental Panel on Climate Change*, Cambridge University Press, p 3.

10. Guy Pearse, *High & Dry: John Howard, Climate Change and the Selling of Australia's Future*, Penguin/Viking, 2007, Endnote 5, p 419
11. IPCC, 2007, "Summary for Policymakers," op cit, p 10.
12. Ibid, p 33.
13. IPCC, *Climate Change 2007: Synthesis Report*, op cit, p 44.
14. Ibid, p 46, figure 3.2.
15. Ibid, p 45.
16. *7:30 Report*, "Scientist Predicts Disastrous Sea Level Rise," ABC TV, 13 March 2007; see also Spratt & Sutton, *Climate Code Red*, op cit.
17. Ross Garnaut, *The Garnaut Climate Change Review, Final Report*, Cambridge University Press, Melbourne, 2008, p 94.
18. R. Schubert, H. J. Schellnhuber, N. Buchmann, A. Epiney, R. Grießhammer, M. Kulessa, D. Messner, S. Rahmstorf and J. Schmid, *Climate Change as a Security Risk*, German Advisory Council on Global Change (WBGU), 2008, p 74.
19. Spratt & Sutton, *Climate Code Red*, op cit, chapter 2.
20. Ross Garnaut, *The Garnaut Climate Change Review, Final Report*, op cit, p 94.
21. IPCC, *Climate Change 2007: Synthesis Report*, op cit, pp 48–49.
22. Stern Review, *The Economics of Climate Change, Part II: The Impacts of Climate Change on Growth and Development*, 2007, p 56; Christian Aid, *Human Tide: The Real Migration Crisis*, 2007, http://www.christian-aid.org.uk.
23. Mrs Edgeworth David, *Funafuti: Or Three Months on a Coral Island: An Unscientific Account of a Scientific Expedition*, John Murray, 1899. See also University of Sydney, Faculty of Science, Edgeworth David (1858–1943), http://www.science.usyd.edu.au/about _us/fame_david.shtml.
24. CIA, *The World Factbook*.
25. Genevieve Sheehan, "Tuvalu Little, Tuvalu Late: A Country Goes Under," *Harvard International Review*, from *International Law*, Vol. 24 (1), Spring 2002, http://hir.harvard.edu/articles/984.
26. R. F. McClean, P. F. Holthus, P. L. Hosking, C. D. Woodroffe, *Tuvalu Land Resources Survey: Vaitupu*, Island Report No 5, Dept of Geography, University of Auckland, 1987, p 28.
27. Guy Pearse, *High & Dry*, op cit, p 213.
28. Christopher C. Horner, "Tuvalunacy," *Washington Post*, 3 April 2002; Patrick Michaels, "Don't Boo-Hoo for Tuvalu," http://www.cato.org, 10 November 2001.

29. N. Mimura, , L. Nurse, R. F. McLean, J. Agard, L. Briguglio, P. Lefale, R. Payet and G. Sem, "Small Islands," in M. L. Parry, O. F. Canziani, J. P. Palutikof, P. J. van der Linden and C. E. Hanson (eds), *Climate Change 2007: Impacts, Adaptation and Vulnerability*, Cambridge University Press, pp 692–93.
30. John Connell, "Losing Ground? Tuvalu, the Greenhouse Effect and the Garbage Can," *Asia Pacific Viewpoint*, Vol 44, no 2, August 2003, pp 89–107.
31. Ibid, p 105.
32. For more information on the denialists see Guy Pearse, *High & Dry*, op cit, and Clive Hamilton, *Scorcher: The Dirty Politics of Climate Change*, Black Inc 2007, particularly chapter 10.
33. "The People of New Zealand," in *Explore Te Ara: The Encyclopedia of New Zealand*, http://www.teara.govt.nz/NewZealandInBrief/Society/1/en.
34. Patrick Michaels, "Don't Boo-Hoo for Tuvalu," op cit.
35. Senator Kerry Nettle, "Campaigns: Climate Change Refugees," available at www.kerrynettle.org.au/300_campaigns_sub.php?deptItemID=51; see also François Gemenne, "Climate Change and Forced Displacements: Towards a Global Environmental Responsibility? The Case of the Small Island Developing States (SIDS) in the South Pacific Ocean," Paper presented to the International Studies Association, San Diego, 22 Mar 2006, http://www.allacademic.com/meta/p98567_index.html.
36. Lianne Dalziel, "Government announces Pacific access scheme," Media Release, 20 December 2001, www.beehive.govt.nz/node/12740.
37. ABC Radio Australia, "Tuvalu ratifies Pacific trade agreement," 28 April 2008, http://www.radioaustralia.net.au/news/stories/200804/s2229643.htm.
38. Sarah Smiles, "Low-skilled labour scheme to be tested this year," *Age*, 18 August 2008, http://www.theage.com.au/national/lowskilled-labour-scheme-to-be-tested-this-year-20080817-3x30.html?page=-1.
39. An internet search of "3000 Tuvaluans" brings up numerous examples.
40. Friends of the Earth, *A Citizen's Guide to Climate Refugees*, 2007, p 8.
41. "South Pacific Peoples, Island Groups and Recent Migration," in *Explore Te Ara: The Encyclopedia of New Zealand*, http://www.teara.govt.nz/NewZealanders/NewZealandPeoples/SouthPacificPeoples/2/en.
42. AusAID & ADB, *Tuvalu: 2006 Economic Report, From Plan to Action*, 2007, Asian Development Bank, Philippines, p 14.
43. Including, for example, Oxfam who published this claim on its website: http://www.oxfam.org.au/campaigns/climate-change/impacts/tuvalu.php.
44. John Connell, "Losing ground?" op cit, pp 99–102.
45. Christopher C. Horner, author of "Tuvalunacy," is a senior fellow at the Competitive Enterprise Institute, http://cei.org.

46. Guy Pearse, *High & Dry*, op cit, pp 201 & 213.
47. AusAID & ADB, *Tuvalu*, op cit, p 14.
48. John Connell, "Losing Ground?" op cit, p 100.
49. Nicholls, R. J., P. P. Wong, V. R. Burkett, J. O. Codignotto, J. E. Hay, R. F. McLean, S. Ragoonaden and C. D. Woodroffe, "Coastal Systems and Low-Lying Areas," in *Climate Change 2007: Impacts, Adaptation and Vulnerability*, op cit, p 321.
50. Pacific Country Report, *Sea Level & Climate: Their Present State, Tuvalu*, June 2006, p 4.
51. Ibid, p 22.
52. Spratt & Sutton, *Climate Code Red*, op cit, p 33.
53. Barrie Macdonald, *Cinderellas of the Empire: Towards a History of Kiribati and Tuvalu*, ANU Press, 1982; Hugh Laracy (ed), *Tuvalu: A History*, Institute of Pacific Studies, 1983.
54. *Tuvalu's National Adaptation Programme of Action*, Ministry of Natural Resources, Environment, Agriculture and Lands, Department of Environment, 2007, pp 35–36.
55. Jane Black, "Tiny Tuvalu Profits From Web Name," *New York Times*, 4 September 2000.
56. This includes revenue estimates of $24.9m plus a Tuvalu Trust Fund distribution in 2008 of A$8.5m (Tuvalu National Budget 2008).
57. François Gemene, "Climate Change and Forced Displacement: Towards Global Responsibility," paper presented at the 47th Annual Convention of the International Studies Association, San Diego, 22–25 March 2006, p 9.
58. Bob Sercombe & Anthony Albanese, *Our Drowning Neighbours: Labor's Policy Discussion Paper on Climate Change in the Pacific*, Australian Labor Party, 2007, p 3.
59. http://en.wikipedia.org/wiki/Kioa
60. Alex Kirby, "Pacific Islanders Flee Rising Seas," BBC News, 9 October 2001, http://news.bbc.co.uk/1/hi/sci/tech/1581457.stm.
61. N. Mimura et al, "Small Islands," op cit, pp 687–716; R. J. Nicholls et al, "Coastal Systems and Low-Lying Areas," pp 315–356.
62. N. Mimura et al, "Small islands," op cit, p 689.
63. Ibid, p 701.
64. Ibid, p 702.
65. 1951 Convention Relating to the Status of Refugees, Article 33
66. Ibid, Article 1A
67. Which recommends that "persons struggling against colonialism" be entitled to protection. See http://www.unhchr.ch/html/menu3/b/o_ asylum.htm.
68. Which includes as grounds for refugee protection "external aggression, occupation, foreign domination or events seriously disturbing public

order." See http://www.africa-union.org/root/au/Documents/Treaties/Text/Refugee_Convention.pdf.

69. Which includes fleeing generalised violence, foreign agresssion, internal conflct and massive violation of human rights as grounds for protection. See http://www1.umn.edu/humanrts/instree/cartagena1984.html.
70. Article 3 of the CAT requires that no state "shall expel, return ('*refouler*') or extradite a person to another State where there are substantial grounds for believing that he [sic] would be in danger of being subjected to torture."
71. The Guiding Principles on Internal Displacement, while not a binding legal instrument, include in its descriptive definition of who might be in need of protection those who are forced to flee their homes – but not cross borders – as a result of "natural or human-made disasters."
72. Megan Saunders, "Ruddock redefining refugees," *Australian*, 22 March 2000, p 5; Andrew Clennell, "Ruddock Pushes UN for Definition of 'Refugee'," *Sydney Morning Herald*, 30 August 2000, p 6. See also Simon Mann, "British Minister Calls for Tighter Refugee Rules, *Age*, 8 February 2001, p 11.
73. See David Marr and Marion Wilkinson, *Dark Victory*, Allen & Unwin, 2nd ed, 2004.
74. See, for example, Micheal Walzer, *Spheres of Justice*, Martin Robertson, 1983.
75. Oli Brown, "Migration and Climate Change," *IOM Migration Research Series*, International Organisation for Migration, No 31, 2008, pp 36–37.
76. UNHCR, 2007 *Global Trends: Refugees, Asylum-seekers, Returnees, Internally Displaced and Stateless Persons*, June 2008, http://www.unhcr.org/statistics/STATISTICS/4852366f2.pdf.
77. Richard Black, "Environmental Refugees: Myth or Reality?" *New Issues in Refugee Research*, Working Paper No 34, March 2001.
78. Stephen Castles, "The International Politics of Forced Migration," *Development*, 43(3), 2003, p 15.
79. Including, among a host of others, El-Hinnawi, *Environmental Refugees,* United Nations Environment Programme, 1985; J. L. Jacobson, *Environmental Refugees: A Yardstick of Habitability,* Worldwatch Institute, 1988.
80. Diane C. Bates, "Environmental Refugees? Classifying Human Migration Caused by Environmental Change," *Population and Environment*, vol 23, no 5, May 2002, pp 465–477.
81. Ibid, p 473.
83. Ibid, p 469.
83. R.J. Nicholls et al, "Coastal Systems and Low-Lying areas," op cit, p 330.
84. Guy Pearse, *High & Dry*, op cit; Clive Hamilton, *Scorcher*, op cit.

85. Ross Garnaut, *The Garnaut Climate Change Review, Final Report*, op cit, p 153.
86. Clive Hamilton, *Scorcher*, op cit, p 93.
87. Ibid, p 94.
88. Pittock's draft text referred to climate change "adding pressure on international aid and immigration policies." In the published book the word "immigration" was replaced with "related," with the same references, including those specifically discussing displacement and migration, cited. (Barrie Pittock, personal communication, 21 August 2008)
89. Senate Hansard, Standing Committee on Legal and Constitutional Affairs, *Estimates*, 21 May 2007.
90. See Peter Mares, *Borderline*, UNSW Press, 2nd Edition, 2002; David Marr and Marion Wilkinson, *Dark Victory*, op cit.
91. Hugh Mackay, "Waking Up Scratchy from the Dreamy Period," *Sydney Morning Herald*, 15 September 2007.
92. David Corlett, "Redefining the Australian Nation," *Eureka Street*, Mar/Apr 2006, http://www.eurekastreet.com.au/article.aspx?aeid=246.
93. Judith Brett, Comment, *The Monthly*, March 2007.
94. Paul Kelly, "Rudd Beyond Kyoto," *Australian*, 19 December 2007, p 10.
95. Michelle Grattan, Mark Forbes and Marian Wilkinson, "Rudd Walks Tightrope on Climate," *Age*, 12 December 2007, p 1.
96. Andrew Bolt, "Global Warmers Set a Hot Pace at Bali," *Herald Sun*, 5 December 2007, p 26.
97. Penelope Debelle, "Dire New Warning on Climate," *Age*, 21 February 2008, p 1.
98. Malcolm Farr, "Heat on Rudd's Climate Pledge - Expert: We Need To Do More, Faster," *Daily Telegraph*, 22 February 2008, p 2.
99. Matthew Franklin, "Nelson Fuels Budget Brawl - Opposition Promises To Slash Petrol Tax, Train Better Teachers," *Australian*, 16 May 2008, p 1.
100. *Canberra Times*, "No Pain, No Gain," 4 July 2008, p 4.
101. Tim Colebatch and Peter Ker, "Nelson Warns Coalition May Not Back Rudd on Climate Change," *Age*, 30 June 2008, p 7.
102. Professor Ross Garnaut, National Press Club Address, 4 July 2008.
103. Robert Manne, Comment, *The Monthly*, August 2008.
104. Samantha Maiden, "MPs Urge Nelson to Lead on Carbon," *Australian*, 9 July 2008, p 6; Michelle Grattan and Penelope Debelle, "Rudd and Howard's Advisers Differ, Nelson Out In the Cold," *Age*, 9 July 2008, p 6; Michelle Grattan, Adam Morton and Chris Hammer, "Liberal Confusion on Emissions Trade Policy," *Age*, 10 July 2008, p 10.
105. Lenore Taylor, "We Must Act Now: Garnaut," *Australian*, 5 July 2008, p 1.
106. Andrew Bolt, "Doomed To a Fatal Delusion," *Herald Sun*, 9 July 2008, p 22.

107. Matthew Warren, "Garnaut's Vision Of Our Green Future," *Australian*, 7 July 2008, p 29.
108. Guy Pearse, *High & Dry*, op cit, p 211.
109. Piers Akerman, "Yes, Minister, It's Just Hot Air," *Sunday Telegraph*, 6 July 2008, p 87.
110. "Garnaut Wisdom Behind the Detail," *Weekend Australian*, 5 July 2008, p 18.
111. http://www.climatechange.gov.au/greenpaper/index.html
112. Ross Garnaut, *The Garnaut Climate Change Review, Final Report*, op cit; *Targets and Trajectories, Supplementary Draft Report*, Garnaut Climate Change Review, 2008, http://www.garnautreport.org.au/reports.
113. Ross Garnaut, *The Garnaut Climate Change Review, Final Report*, op cit, ch 12.
114. Lenore Taylor, "Garnaut Pushes Low-Key Target," *Australian*, 6 September 2008, p.1; Adam Morton and Chris Hammer, "Garnaut Takes 'Softly Softly' Line On Cutting Greenhouse Emissions," *Age*, 6 September 2008, p 1.
115. 'Softly, Softly on Garnaut Report,' *Canberra Times*, 1 October 2008, p 10.
116. Rosslyn Beeby, "Garnaut Urges 90pc Cut by 2050," *Canberra Times*, 1 October 2008, p 2.
117. "No Pain, No Gain," *Canberra Times*, 4 July 2008, p 4; Lenore Taylor, "Action on Climate Change Can't Wait, Voters Say," *Australian*, 15 July 2008; Michelle Grattan, "Big Tick for Emissions Cuts," *Age*, 21 July 2008.
118. Bob Sercombe & Anthony Albanese, *Our Drowning Neighbours: Labor's Policy Discussion Paper on Climate Change in the Pacific*, Australian Labor Party, 2007, p 4.
119. Ibid, p 5.
120. Ibid, p 10.
121. Bob McMullan, Duncan Kerr, "Australia to Help Pacific Countries Tackle Climate Change," Media Release, 3 February 2008.
122. Lloyd Jones, "PAC: Pacific Looks to Softer Rudd After 'Tough Cop' Howard Years," AAP, 14 December 2007.
123. Senate Hansard, Standing Committee on Legal and Constitutional Affairs, *Estimates*, 19 February 2008, p 56.
124. Senate Hansard, Standing Committee on Legal and Constitutional Affairs, 28 May 2008, p 31.
125. The Hon Stephen Smith MP, Minister for Foreign Affairs, The Hon Bob McMullan MP, Parliamentary Secretary for International Development Assistance, The Hon Duncan Kerr MP, Parliamentary Secretary for Pacific Island Affairs, "2008–09 International Development Assistance Budget," http://www.bobmcmullan.com.

126. Personal communication with Simon Dowding, media adviser to Senator Chris Evans, Minister for Immigration and Citizenship, 17 July 2008.
127. Telephone conversation with Simon Dowding, media adviser to Senator Chris Evans, 2 September 2008.
128. Senate Hansard, Standing Committee on Legal and Constitutional Affairs, 28 May 2008, p 30.
129. Dilip Ratha, "Global Remittances," interviewed on by Philip Adams on Late Night Live, 15 July 2008.
130. Oli Brown, *Migration and Climate Change*, op cit, p 39.
131. Jock Collins, *Migrant Hands in a Distant Land: Australia's Post-War Immigration*, Pluto Press, 1991; Paul Bartrop, *Australia and the Holocaust 1933–45*, Australian Scholarly Publishing, 1994.
132. Hansard, House of Representatives, 24 May 1977, p 1714.
133. Klaus Neumann, *Refuge Australia: Australia's Humanitarian Record*, UNSW Press, 2004, pp 79–91.
134. Nancy Viviani, *The Long Journey: Vietnamese Migration and Settlement in Australia*, MUP, Carlton, 1984; James Hathaway, "Labelling the 'Boat People': The Failure of the Human Rights Mandate of the Comprehensive Plan of Action for Indochinese Refugees," Human Rights Quarterly 15, 1993, pp 686–702.
135. Ross Garnaut, *The Garnaut Climate Change Review, Final Report*, op cit, p 42–43.
136. Ibid p 45.
137. Ibid, p 89.
138. Ibid, p 46.
139. David Spratt & Philip Sutton, op cit, ch.1.
140. Ibid, p 17.
141. Ibid, p 33.
142. Department of Immigration and Citizenship, "Fact Sheet 2 – Key Facts in Immigration," http://www.immi.gov.au/fact-sheets/02key.htm.
143. Department of Immigration and Citizenship, "Fact Sheet 60 – Australia's Refugee and Humanitarian Program," http://www.immi.gov.au/media/fact-sheets/60refugee.htm.
144. Although the Carteret Islanders have already begun relocating to Bougainville, the reasons for their homeland becoming uninhabitable are somewhat controversial. There have been suggestions that locals themselves damaged the reefs that protected their land from the sea by dynamite-fishing during the Bougainville war (Greg Roberts, "Islanders Face Rising Sea With Nowhere To Go," *Sydney Morning Herald*, 30 March 2002). Others have suggested that sea inundation may be due to the subsidence of the land (Richard Lloyd Parry, "The Last Tide Could Come At Any Time," *Times*, 21 December 2006, p 40). The case of the

Carteret Islanders may exemplify the complex dynamics between local and global environmental factors involved in climate-related dislocation.

145. Rafael Reuveny, "Ecomigration and Violent Conflict: Case Studies and Public Policy Implications," *Hum Ecol* (2008) 36:1–13, p 9.
146. R. Schubert, H. J. Schellnhuber, N. Buchmann, A. Epiney, R. Grießhammer, M. Kulessa, D. Messner, S. Rahmstorf, J. Schmid, *Climate Change as a Security Risk*, op cit, pp 59, 124–127, 136–138.
147. UNDP, 2007, *Human Development Report 2007/2008, Fighting Climate Change: Human Solidarity in a Divided World*, UNDP, p 77.
148. Julian Borger, "Conflicts fuelled by climate change causing new refugee crisis, warns UN," *Guardian*, 17 June 2008.
149. IPCC, *Special Report: Emissions Scenarios*, "Summary for Policy Makers," IPCC 2000.
150. Ross Garnaut, *The Garnaut Climate Change Review, Final Report*, op cit, p 86.
151. R. Schubert et al, op cit, p 78.
152. Ibid, pp 122–124.
153. World Bank, "Gross National Income Per Capita 2007, Atlas Method and PPP," http://siteresources.worldbank.org/DATASTATISTICS/Resources/GNIPC.pdf.
154. Robert Manne and David Corlett, *Sending Them Home: Refugees and the New Politics of Indifference*, Black Inc, 2004.
155. PNG was part of the Pacific Solution. In 2004 it was announced that Italy was planning to build an "immigrant reception centre" in Libya where asylum seekers could be processed outside the EU (Benedetto Cataldi, "Analysis: Italy Woos Libya," *BBC News*). Romania, along with Albania, Croatia, Iran, Morocco, northern Somalia, Russia, Turkey and Ukraine, was suggested as a place where asylum seekers attempting to get to Europe could be kept and have their claims processed extraterritorially. (Human Rights Watch, *An Unjust "Vision" for Europe's Refugees*, 17 June 2003, www.hrw.org/backgrounder/refugees/uk/newvision.pdf.)
156. Anthony Richmond, *Global Apartheid: Refugees, Racism and the New World Order*, Oxford University Press, 1994.
157. James C. Hathaway & R. Alexander Neve, "Making International Refugee Law Relevant Again: A Proposal for Collectivized and Solution-Oriented Protection," *Harvard Human Rights Journal*, vol 10, 1997, pp 115–211; Peter Schuck, "Refugee Burden-Sharing: A Modest Proposal," *Yale Journal of International Law*, vol 22, no 2, 1997, pp 243–297. See, in response, Deborah Anker, Joan Fitzpatrick and Andrew Shacknove, "Crisis and Cure: A Reply to Hathaway/Neve and Schuck," *Harvard Human Rights Journal*, vol.11, 1998, pp 295–310.

MIX
Papier | Fördert gute Waldnutzung
FSC® C083411

Zeitfracht Medien GmbH
Ferdinand-Jühlke-Straße 7
99095 Erfurt, Deutschland
produktsicherheit@kolibri360.de